NoOps

How AI Agents Are Reinventing DevOps and Software

Roman Vorel

apress®

NoOps: How AI Agents Are Reinventing DevOps and Software

Roman Vorel
Weddington, NC, USA

ISBN-13 (pbk): 979-8-8688-1693-2 ISBN-13 (electronic): 979-8-8688-1694-9
https://doi.org/10.1007/979-8-8688-1694-9

Copyright © 2025 by Roman Vorel

This work is subject to copyright. All rights are reserved by the Publisher, whether the whole or part of the material is concerned, specifically the rights of translation, reprinting, reuse of illustrations, recitation, broadcasting, reproduction on microfilms or in any other physical way, and transmission or information storage and retrieval, electronic adaptation, computer software, or by similar or dissimilar methodology now known or hereafter developed.

Trademarked names, logos, and images may appear in this book. Rather than use a trademark symbol with every occurrence of a trademarked name, logo, or image we use the names, logos, and images only in an editorial fashion and to the benefit of the trademark owner, with no intention of infringement of the trademark.

The use in this publication of trade names, trademarks, service marks, and similar terms, even if they are not identified as such, is not to be taken as an expression of opinion as to whether or not they are subject to proprietary rights.

While the advice and information in this book are believed to be true and accurate at the date of publication, neither the authors nor the editors nor the publisher can accept any legal responsibility for any errors or omissions that may be made. The publisher makes no warranty, express or implied, with respect to the material contained herein.

    Managing Director, Apress Media LLC: Welmoed Spahr
    Acquisitions Editor: Celestin Suresh John
    Development Editor: Jim Markham
    Coordinating Editor: Gryffin Winkler

Cover image designed by Freepik (www.freepik.com)

Distributed to the book trade worldwide by Springer Science+Business Media New York, 1 New York Plaza, New York, NY 10004. Phone 1-800-SPRINGER, fax (201) 348-4505, e-mail orders-ny@springer-sbm.com, or visit www.springeronline.com. Apress Media, LLC is a Delaware LLC and the sole member (owner) is Springer Science + Business Media Finance Inc (SSBM Finance Inc). SSBM Finance Inc is a **Delaware** corporation.

For information on translations, please e-mail booktranslations@springernature.com; for reprint, paperback, or audio rights, please e-mail bookpermissions@springernature.com.

Apress titles may be purchased in bulk for academic, corporate, or promotional use. eBook versions and licenses are also available for most titles. For more information, reference our Print and eBook Bulk Sales web page at http://www.apress.com/bulk-sales.

Any source code or other supplementary material referenced by the author in this book can be found here: https://www.apress.com/gp/services/source-code.

If disposing of this product, please recycle the paper

To my wife—your steady faith turned late-night ideas into daylight possibilities.

To our two extraordinary sons—David and Damian—your boundless curiosity and unfiltered questions remind me why we build the future, not just for efficiency but for wonder.

This book is for you, with all my love and gratitude.

Table of Contents

About the Author xix

Preface xxi

Part I: Standardization, Cloud-Native and Data-Driven DevOps 1

Chapter 1: The Evolution of DevOps 3

1.1 From Silos to Collaboration 3
    1.1.1 The Traditional Divide 3
    1.1.2 The Agile Roots 4

1.2 Early Pioneers and Defining Moments 5
    1.2.1 Patrick Debois and the "DevOps" Term 5
    1.2.2 The Phoenix Project Influence 5

1.3 DevOps Core Principles 6

1.4 Success Stories and the Promise of DevOps 7
    1.4.1 High-Performing Organizations 7
    1.4.2 Key Measurable Benefits 8

1.5 New Pressures and Emerging Challenges 9

1.6 Toward an Expanded Vision: DevSecOps, DataOps, and NoOps 11
    1.6.1 From DevOps to DevSecOps 11
    1.6.2 DataOps, MLOps, etc. 11
    1.6.3 The Rise of "NoOps" 12

1.7 DevOps Meets AI: A Glimpse Ahead 12

TABLE OF CONTENTS

 1.8 Change-Management Frameworks for an AI-Driven DevOps Journey 13
  1.8.1 Why Change Management Is Nonoptional 14
  1.8.2 Classic Frameworks and Their Fit for AI-DevOps 14
  1.8.3 A Hybrid Playbook—A-DAIR for AI-DevOps 15
  1.8.4 Embedding Change Management in the DevOps Loop 16
  1.8.5 Quick-Start Checklist 16
  1.8.6 Key Takeaways 17
 1.9 Chapter Summary and Looking Ahead 17
 1.10 Key Takeaways 19

Chapter 2: Fragmented Software Development: Why DevOps Isn't Always Enough 21

 2.1 The Rise of Tool Sprawl 22
  2.1.1 The Allure of Specialized Tools 22
  2.1.2 Number of Tools and the "Tool Tax" 23
 2.2 Data Silos and Lack of End-to-End Visibility 24
  2.2.1 Fragmented Data Landscape 24
  2.2.2 The Visibility Gap 25
 2.3 Impact on Collaboration and Workflow 25
  2.3.1 DevOps Irony: New Silos 25
  2.3.2 Collaboration Friction and Context Switching 26
 2.4 The Culture of "Choose Your Own Tool" 27
 2.5 The Hidden Costs of Fragmentation 28
  2.5.1 Slowed Time to Market 28
  2.5.2 Increased Risk of Errors 28
  2.5.3 Lower Morale and Higher Burnout 28
  2.5.4 Difficulty Scaling 29

2.6 Real-World Example: A Financial Services Firm in "Tool Chaos" 29
    2.6.1 Multiple CI/CD Tools, Repos, and Scripts 29
    2.6.2 The Complexity of Multiple IDEs 30
2.7 The AI Readiness Angle 32
2.8 Why Fragmentation Persists 32
2.9 The Way Forward 33
    2.9.1 Recognize the Cost 33
    2.9.2 Plan for Standardization 33
    2.9.3 Evolve from DevOps to Platform Engineering 33
    2.9.4 Focus on Data Centralization 34
    2.9.5 Standardize the Developer Experience 34
2.10 Chapter Summary and Looking Ahead 35
2.11 Key Takeaways 36

Chapter 3: The Case for Standardization: Building the Foundation for NoOps 39

3.1 What Do We Mean by "Standardization"? 40
    3.1.1 Defining Standardization in DevOps 40
    3.1.2 Why It Matters More Than Ever 41
3.2 The Core Benefits of Standardization 41
    3.2.1 Streamlined Collaboration 41
    3.2.2 Reduced Operational Overhead (the Anti-Tool-Tax) 42
    3.2.3 Stronger Security and Compliance 42
    3.2.4 Increased AI Readiness 43
3.3 Addressing Fears and Misconceptions 43
    3.3.1 "Won't Standardization Kill Innovation?" 43
    3.3.2 "It's Too Hard to Switch from Existing Tools" 44
    3.3.3 "We Need Different Tools for Different Languages or Frameworks" 44

TABLE OF CONTENTS

 3.4 Approaches to Standardization 45
  3.4.1 Platform Engineering and the Internal Developer Platform 45
  3.4.2 Reference Architectures and Golden Pipelines 45
  3.4.3 Standardizing the Developer Experience 46
  3.4.4 Data Unification 47
 3.5 Standardization As the Launchpad for AI 47
  3.5.1 AI Demands High-Quality Data 47
  3.5.2 Enabling Autonomous Agents 48
 3.6 Case Study: A Global Tech Firm's "Platform First" Approach 48
 3.7 Chapter Summary and Looking Ahead 50
 3.8 Key Takeaways 50

Chapter 4: Cloud-Native and Data-Centric Approaches 53
 4.1 Why "Cloud-Native" Matters 53
  4.1.1 Definition and Core Principles 53
  4.1.2 The Shift from Monoliths to Microservices 54
 4.2 Containerization and Ephemeral Infrastructure 55
  4.2.1 Containers vs. Virtual Machines 55
  4.2.2 Orchestration with Kubernetes 56
 4.3 Infrastructure as Code (IaC) 57
  4.3.1 Principles of IaC 57
  4.3.2 Popular IaC Tools 57
  4.3.3 Why IaC Complements DevOps 58
 4.4 Data-Centric Architectures and Observability 58
  4.4.1 Breaking Down Siloed Data 58
  4.4.2 Observability vs. Monitoring 59
  4.4.3 Real-Time Feedback Loops 60
  4.4.4 Platform-Agnostic Analytics 60

TABLE OF CONTENTS

4.5 Putting It All Together: Integrated Cloud-Native Pipelines............................61
    4.5.1 A Typical Workflow..61
    4.5.2 Security and Compliance in the Pipeline...............................62
4.6 Case Study: Retail Giant Embracing Cloud-Native 63
4.7 Why This Matters for AI and NoOps..64
    4.7.1 Cloud-Native + Standardization = Data Gold Mine..............64
    4.7.2 Autonomous Scaling and Self-Healing 65
    4.7.3 Rapid Adoption of New AI Capabilities.................................65
    4.7.4 Developer in the Loop…For Now 66
4.8 Key Takeaways and Next Steps 66
    4.8.1 What's Next?..68
4.9 Chapter Summary...68

Chapter 5: What "Good" Looks Like: A Reference Architecture..........71

5.1 The Pillars of a "Good" DevOps Architecture......................................72
    5.1.1 End-to-End Integration 72
    5.1.2 A Single Source of (Structured) Data....................................72
    5.1.3 Self-Service and Self-Healing 73
    5.1.4 Embedded Security and Compliance....................................73
5.2 Reference Model Overview 74
5.3 Example Workflow in Action..76
5.4 Organizational Design: The Supporting Structure 79
5.5 Hallmarks of a Mature Reference Architecture...................................80
5.6 Real-World Example: A SaaS Company's Unified Pipeline 82
5.7 Common Pitfalls and How to Avoid Them 83
5.8 The Road Ahead 85
5.9 Chapter Summary...85

TABLE OF CONTENTS

5.10 Final Section (Part I): The Paved Road—Standardization, Cloud-Native Foundations, and Unified Insights 87

5.11 Executive Snapshot..87

5.12 Key Takeaways..88

5.13 Common Pitfalls...90

5.14 Mitigation Playbook—From Strategy to Daily Habit 92

5.15 Implementation Guidance—Turning the Vision into an Org-Wide Upgrade Path 94

    5.15.1 Quick-Start Checklist...95

    5.15.2 Sequenced Migration Plan 96

    5.15.3 KPIs and Success Metrics (All Surfaced in Opsera)................98

5.16 Glossary—Part I...99

Part II: Generative AI Transformations..................................103

Chapter 6: Generative AI for Coding and Unit Testing..................105

6.1 The Rise of AI Coding Assistants 105

    6.1.1 From Autocomplete to Intelligent Pair Programming 105

    6.1.2 Why This Is a Game-Changer.......................................106

6.2 Generative AI in Practice: Coding Workflows 108

    6.2.1 Prompting and Refining with GitHub Copilot 108

    6.2.2 Handling Edge Cases and Documentation............................108

    6.2.3 Team Collaboration and Code Reviews..............................109

6.3 Impact on Productivity and Code Quality 109

6.4 AI-Driven Unit Test Generation 111

    6.4.1 Why Automated Test Creation?.....................................111

    6.4.2 Example Workflow with GitHub Copilot 112

    6.4.3 Benefits and Limitations...112

6.5 Challenges and Limitations of Generative AI in Coding...............113

TABLE OF CONTENTS

6.6 Best Practices for AI Coding and Unit Testing 114

6.7 The Road Toward Advanced AI-Driven Development 115

    6.7.1 Evolution of Code Suggestions 115

    6.7.2 Unified Developer Experience 116

    6.7.3 Bridging to NoOps 116

6.8 Chapter Summary 116

Chapter 7: Generative AI for System and Integration Testing 119

7.1 Why Functional and Integration Testing Matter 120

    7.1.1 From Unit Tests to Real-World Scenarios 120

    7.1.2 The Pain of Manual Test Maintenance 120

7.2 The Rise of AI-Driven Functional Testing 121

    7.2.1 Functionaize As a Prime Example 121

    7.2.2 AI-Powered End-to-End Validation 122

7.3 AI-Enhanced Testing Workflows 122

    7.3.1 Generating Tests 122

    7.3.2 Self-Healing in Action 123

    7.3.3 Integration Testing Across Services 123

7.4 Benefits and Limitations of AI-Driven Functional Testing 124

    7.4.1 Key Benefits 124

    7.4.2 Challenges and Caveats 124

7.5 Best Practices for Incorporating AI-Based Functional Testing 125

7.6 Case Study: Ecommerce Platform Adopting Functionaize 126

7.7 The Road Ahead: AI Testing and the NoOps Vision 128

    7.7.1 Beyond Scripts: Autonomous Test Agents 128

    7.7.2 Closing the Gap Between Dev, QA, and Ops 128

    7.7.3 OpenAI Operator: A Glimpse of Future System Testing 129

7.8 Chapter Summary 130

TABLE OF CONTENTS

Chapter 8: Generative AI for IaC and Data Provisioning 133

8.1 Why AI for IaC and Data Provisioning? 134

    8.1.1 Complexity and Rapid Changes 134

    8.1.2 Seamless Integration with DevOps and IDE NLP 135

8.2 AI-Driven IaC Generation and Data Provisioning 136

    8.2.1 Automated Script Creation 137

    8.2.2 Refactoring and Modernization 138

8.3 Predictive Scaling, Drift Remediation, and Data Refresh 138

    8.3.1 Predictive Scaling 138

    8.3.2 Drift and Misconfiguration Remediation 139

8.4 "Stay in the Flow": IDE-Centric, NLP-Driven Actions 140

8.5 Best Practices for AI-Driven IaC and Data Management 141

8.6 Case Study: AI-Assisted Terraform and Data Masking at a FinTech Startup 142

8.7 The Road Ahead: Self-Healing Infrastructure and Data, Stay-in-Flow Approach 144

    8.7.1 Multiagent Infrastructure and Data Management 144

    8.7.2 Operator-like Autonomy in Infrastructure and Data 145

    8.7.3 NLP-Driven Flow in the IDE 145

    8.7.4 Toward NoOps 146

8.8 Chapter Summary 146

Chapter 9: AI-Orchestrated CI/CD and Pipeline Optimization 149

9.1 The Need for Smarter Pipelines 150

    9.1.1 Complexity and Staging Bottlenecks 150

    9.1.2 Real-Time Feedback vs. Blind Scripts 151

9.2 AI-Driven Pipeline Optimization 152

    9.2.1 Intelligent Test Selection 152

    9.2.2 Partial/On-Demand Deployment Sequences 153

TABLE OF CONTENTS

9.3 Predictive Failure Analysis and Remediation 154
    9.3.1 Anomaly Detection 154
    9.3.2 Auto-Apply Fixes or Reruns 154

9.4 Deploy and Release Strategy Optimization 155
    9.4.1 Blue-Green, Canary, and Rolling 155
    9.4.2 Real-Time Telemetry Feedback 155

9.5 Stay in the Flow: IDE-Centric, NLP-Driven CI/CD Control 156
    9.5.1 Natural Language Triggers 156
    9.5.2 Quick Feedback and Reduced Context Switching 156

9.6 Best Practices for AI-Orchestrated CI/CD 157

9.7 Case Study: Ecommerce Company's AI-Managed Pipeline 158

9.8 The Road Ahead: AI Pipeline Agents and NoOps 160
    9.8.1 Multiagent Pipeline Collaboration 160
    9.8.2 Real-Time Observations and Automated Fixes 160
    9.8.3 NLP-Driven Flow from IDE 161

9.9 Chapter Summary 161

9.10 Final Section (Part II): Catalyst to Autonomy—Generative AI
      Foundations for the Multiagent NoOps Era 163

9.11 Executive Snapshot 163

9.12 Key Takeaways 166

9.13 Common Pitfalls 167

9.14 Mitigation Playbook—Hardening AI from Experiment to
      Everyday Muscle Memory 167
    9.14.1 Platform Guardrails 168
    9.14.2 AI-Assisted Coding and Testing 169
    9.14.3 Infrastructure and Operations 170
    9.14.4 Security and Compliance 171
    9.14.5 Adoption and Business KPIs (All via Opsera) 172

TABLE OF CONTENTS

9.15 Implementation Guidance and Checklist—Turning AI Ambition into a Measurable Rollout 173

    9.15.1 Quick-Start Checklist..................173

    9.15.2 Sequenced Migration Plan 175

    9.15.3 KPIs and Success Metrics (All via Opsera Unified Insights) 177

9.16 Glossary—Part II..................179

Part III: Multiagent AI and the NoOps Horizon 181

Chapter 10: Autonomous Multiagent Systems 183

10.1 Beyond Single AI Tools: The Multiagent Synergy..................184

    10.1.1 A Team of AI Specialists..................184

    10.1.2 Communication and Decision-Making..................185

10.2 The Path to Autonomous NoOps..................186

    10.2.1 Fewer Manual Touchpoints..................186

    10.2.2 Intelligent Collaboration..................186

10.3 Benefits and Challenges of Multiagent Systems..................187

    10.3.1 Key Benefits 187

    10.3.2 Challenges 188

10.4 Real-World Example: Toward an Integrated AI-Powered DevOps..................189

10.5 NLP and IDE Integration: "Stay in the Flow" for Everything..................190

    10.5.1 Unified Interface 190

    10.5.2 Minimal Context Switching..................190

10.6 Best Practices for Embracing Multiagent NoOps 191

10.7 Looking Forward: The Emerging NoOps World..................192

    10.7.1 Ultimate State of Autonomy..................192

    10.7.2 Continued Role for Humans..................193

    10.7.3 Constant Evolution..................193

10.8 Chapter Summary 194

TABLE OF CONTENTS

Chapter 11: Human–AI Collaboration 197

11.1 The Shifting Role of Humans in a NoOps Landscape 198
11.1.1 From Manual Operators to Automation Architects 198
11.1.2 Developers As Product Creators 199
11.1.3 QA As Quality Engineers 199
11.1.4 Security and Compliance Roles 200

11.2 Building Trust in AI 200
11.2.1 Human-in-the-Loop Approach 200
11.2.2 Auditable Actions and Policy Checks 201
11.2.3 Transparency and Explainability 202

11.3 Upskilling and Team Dynamics 202
11.3.1 Training Developers and Ops 202
11.3.2 Collaboration with AI Agents 203
11.3.3 New Roles and Leaner Teams 203

11.4 Ethical and Compliance Considerations 204
11.4.1 Boundaries of AI Autonomy 204
11.4.2 Bias and Reliability 204
11.4.3 Legal and Accountability 205

11.5 Cultural and Organizational Shifts 205
11.5.1 Embracing AI As a Teammate 205
11.5.2 Learning from Failures 206
11.5.3 Continuous Iteration on Roles and Processes 206

11.6 The Long-Term NoOps Vision 206
11.6.1 Humans As Strategic Overseers 206
11.6.2 Lifelong Learning and Evolving AI 207
11.6.3 The Human Touch 207

11.7 Chapter Summary 208

TABLE OF CONTENTS

Chapter 12: The Future of Software Development 211

12.1 From DevOps to NoOps—What's Next? 212
  12.1.1 Full Lifecycle AI...212
  12.1.2 Multiagent Collaboration at Scale 213
12.2 Autonomous Code Generation and Live Agentic Collaboration 213
  12.2.1 Code As Conversation..213
  12.2.2 Interactive Agents in the IDE 214
12.3 The Workforce and Organizational Impact 215
  12.3.1 Upskilling and New Roles..215
  12.3.2 Leaner Teams, Faster Delivery 215
12.4 Business Strategy and Competitive Advantage 216
  12.4.1 Time to Market and Continuous Innovation 216
  12.4.2 Data Monetization and AI Feedback Loops..216
12.5 Challenges and Limitations in the Emerging NoOps Era 217
  12.5.1 Complexity and Interagent Conflicts..217
  12.5.2 Ethical and Legal Hurdles 217
  12.5.3 Trust and Cultural Adoption 218
12.6 Is Complete NoOps Truly Attainable? 218
  12.6.1 The Last Mile of Human Judgment..218
  12.6.2 The Ongoing Collaboration..219
12.7 Chapter Summary and Conclusion..219
12.8 Conclusion and Final Thoughts 221
12.9 Recap of the Journey 221
12.10 Why It Matters Now..223
12.11 A Practical Call to Action 224
12.12 The Ongoing Evolution 225

12.13 Final Reflections	226
12.14 Glossary—Part III	227
12.15 Bibliography	228
Index	**231**

About the Author

Roman Vorel is a visionary technologist and transformative leader who bridges the gap between cutting-edge innovation and human-centered collaboration. A graduate of Brno University of Technology, where he earned a Master of Science in Computer Science and Engineering, and Nottingham Trent University, where he honed his strategic leadership skills with an MBA in Business Management, Roman's academic foundation mirrors his career ethos: blending technical rigor with purposeful execution.

With a career dedicated to redefining how global enterprises harness technology, Roman has held pivotal leadership roles at Honeywell, a Fortune 100 industrial technology pioneer. As Corporate Vice President and Chief Information Officer (CIO), he spearheaded the modernization of software development for thousands of engineers, embedding AI into every phase of the innovation lifecycle. Earlier, he led global supply chain and digital transformation initiatives, unifying fragmented systems into cohesive cloud architectures and replacing siloed workflows with real-time visibility. His tenure also included driving enterprise-wide ERP consolidations as a Global Deployment Leader, reshaping IT operations across continents, and fostering collaboration in regions spanning the Americas, EMEA, and APAC.

What truly sets Roman apart is his belief that technology's greatest power lies in its ability to elevate human potential. By marrying technical precision with empathy, he has proven that even in Fortune 100

ABOUT THE AUTHOR

environments, innovation thrives when people are empowered, processes are democratized, and data becomes a universal language for progress. Whether architecting AI-driven analytics, mentoring the next generation of leaders, or bridging ambition with execution, Roman's work centers on creating systems that amplify collective ingenuity.

Today, as a thought leader and advocate for AI-driven collaboration, Roman continues to inspire organizations to rethink what's possible—not just in engineering, but in how we align technology with humanity's most pressing challenges. His story is a testament to the idea that the future belongs to those who dare to unify vision with action, recognizing that transformative systems are those built by—and for—people.

Preface

Software is now the heartbeat of every industry, beating faster with each new release, feature flag, and security patch. Yet many teams still spend most of their day chasing down flaky tests, wrestling with sprawling pipelines, and firefighting midnight outages. It's no surprise that the original promise of DevOps—"build, ship, learn, repeat"—often feels more like "build, ship, burn out."

Meanwhile, a new force has arrived at full volume: generative AI. Large language models can draft code, design self-healing tests, write Terraform, correlate logs, and even decide when and how to roll back a risky deployment. When these capabilities are wired into a disciplined DevOps platform, the result is something far closer to **NoOps**—a state in which the drudgery of day-to-day operations melts away and engineers are free to create.

This book is your field guide to that future. It is **not** a hand-wavy ode to "AI magic," nor a collection of disconnected hacks. Instead, it offers a play-by-play blueprint:

- **Part I** shows how to eliminate toolchain chaos and standardize on cloud-native, data-centric foundations—the nonnegotiable launchpad for any serious AI initiative.

- **Part II** layers in generative AI, step by step: coding assistants that raise velocity, self-healing functional tests that slash QA overhead, infrastructure agents that prevent drift, and pipeline bots that run canaries, rollbacks, and compliance gates while you sleep.

- **Part III** looks over the horizon at autonomous multiagent systems—and explains how to keep humans in charge of ethics, strategy, and innovation as NoOps becomes real.

Along the way, you'll find candid war stories, measurable KPIs, and ready-to-run playbooks you can lift into your own organization tomorrow morning.

If you're an engineering leader tired of watching cycle times stagnate; a platform or SRE veteran drowning in alert fatigue; a developer who'd rather craft features than copy-paste boilerplate; or an executive betting your business on faster, safer digital delivery—this book will show you how AI can turn DevOps into a competitive weapon instead of a cost center.

The road ahead is bold, occasionally bumpy, but undeniably exciting. Let's take the first step toward a world where software almost runs itself—and people get back to the creative work only people can do. Welcome to **NoOps Nation**.

Who Is This Book For

- **Technical Executives and Managers**

    For directors and VPs aiming to boost developer productivity or drive a wide-scale digital transformation, these chapters demonstrate how to layer AI into DevOps strategies, bridging technology and business outcomes.

- **DevOps Practitioners**

    If you already embrace DevOps principles but struggle with fragmented toolchains, manual test overhead, or slow release cycles, this book offers a practical roadmap to streamline and enhance your workflow with AI's help.

- **Engineering and QA Leads**

    Leaders who oversee application development or testing teams can learn how to automate code suggestions, self-heal functional tests, and unify data provisioning—all in a unified, AI-driven manner.

- **Ops and SRE Professionals**

    If you're responsible for uptime, infrastructure, or production incidents, see how AI-based provisioning, predictive scaling, and drift remediation can reduce firefighting and deliver more stability.

- **Curious Developers**

    Even if you're new to DevOps or AI, you'll find step-by-step guidance on integrating generative models into your day-to-day coding, testing, and CI/CD routines—removing repetitive tasks and accelerating feedback loops.

- **Visionaries Envisioning NoOps**

    Those eager for a glimpse of software delivery's future—where manual toil is minimal—will find both inspiration and cautionary advice on how to balance automation with the oversight and creativity only humans can provide.

No matter your role, if your goal is to modernize software delivery while unleashing AI's potential for better, faster releases, the insights in this book are for you.

PART I

Standardization, Cloud-Native and Data-Driven DevOps

"Every meaningful transformation starts by exposing the friction that slows us down. Standardization removes that drag; cloud-native, data-centric practices propel us forward—making our pipelines elastic, observable, and ready for continuous change."

CHAPTER 1

The Evolution of DevOps

The emergence of DevOps in the late 2000s marked a transformative moment in how organizations develop, deliver, and maintain software. It began as a cultural shift—aimed at knocking down the wall between development (Dev) and operations (Ops)—and it has evolved into a robust set of practices, tools, and mindsets that push software delivery to be faster, more reliable, and more secure. Yet, like any major movement, DevOps did not arise in isolation. It was shaped by the frustrations of siloed teams, the rise of Agile methodologies, and the increasing customer demand for always-available, continuously updated digital products. This chapter explores **why** DevOps emerged, **how** it revolutionized software delivery, and **where** it still falls short in today's complex, rapidly changing tech landscape—setting the stage for the transformations examined throughout this book.

1.1 From Silos to Collaboration

1.1.1 The Traditional Divide

In the traditional model of software development, teams were rigidly split by function. Developers wrote code in a vacuum, often with minimal insight into how the software would actually run in production.

CHAPTER 1   THE EVOLUTION OF DEVOPS

After months (or even years) of coding, a "finished" product would be thrown over the wall to operations, whose job was to deploy and manage it on physical servers. If performance issues, bugs, or downtime arose, operations had to scramble to diagnose them—while developers, having moved on to the next project, were rarely on the hook for fixes.

This model suffered from

- **Long Release Cycles**: Major releases sometimes happened only once or twice a year, or even less frequently.
- **Blame Culture**: When production issues surfaced, dev and ops teams often pointed fingers at each other.
- **Siloed Knowledge**: Developers knew code but not production environments; operations knew environments but not application logic.
- **Lack of Feedback Loops**: Developers received little insight into how users interacted with their software once it was deployed, limiting opportunities to improve quickly.

The resulting friction slowed time to market, hurt software quality, and caused frustration across the organization. These dynamics set the stage for a more collaborative approach.

1.1.2 The Agile Roots

In parallel with these frustrations, **Agile methodologies** like Scrum and Extreme Programming (XP) gained popularity in the early 2000s. Agile emphasized **short development sprints**, frequent feedback, and close collaboration with stakeholders. However, while Agile addressed many issues in the development cycle (faster iteration, user-centric design), operations teams were still largely outside this iterative loop. Agile projects would still hit a bottleneck at deployment time—where the iteration "stopped" and the old dev-ops divide reasserted itself.

As software teams embraced faster iteration, the need for an equally rapid, **continuous** approach to deployment and infrastructure management grew. This was the gap DevOps sought to fill: bridging the principles of Agile with the realities of running software at scale.

1.2 Early Pioneers and Defining Moments

1.2.1 Patrick Debois and the "DevOps" Term

Many credit **Patrick Debois**, a Belgian consultant, as one of the earliest champions of DevOps. Around 2007, Debois recognized a recurring clash between development and operations in Agile projects. Seeking solutions, he helped organize "DevOpsDays" conferences starting in 2009, which became a grassroots movement that rapidly grew through meetups, open conferences, and an enthusiastic online community.

The term "DevOps" itself emerged around this time—specifically tied to the 2009 **Velocity Conference** talk by John Allspaw and Paul Hammond titled "10+ Deploys Per Day: Dev and Ops Cooperation at Flickr." Their presentation showcased how Flickr's engineering team had broken traditional release cycles by deploying code multiple times per day, while working hand in hand with operations. This was a eureka moment for many engineers who realized: if Flickr could do that, maybe their companies could too.

1.2.2 The Phoenix Project Influence

Another milestone in popularizing DevOps was the novel *The Phoenix Project* by Gene Kim, Kevin Behr, and George Spafford, published in 2013. Presented as a story, it illustrated common dysfunctions—endless firefighting, siloed teams, management blind spots—and how adopting collaborative, automated, and flow-oriented practices could turn a struggling IT department into a strategic advantage for the business. The success of that book introduced DevOps ideas to mainstream corporate leadership and turned more eyes to continuous integration (CI), continuous delivery (CD), and a "culture of shared responsibility."

1.3 DevOps Core Principles

Although DevOps can be interpreted in many ways, there are some foundational themes that nearly every DevOps initiative aims to uphold:

1. **Collaboration and Shared Responsibility**
   DevOps breaks down walls. Developers and operations (and more recently, security) share accountability for the software's performance, uptime, and user experience. If the production site goes down, dev and ops solve it together rather than pointing fingers.

2. **Continuous Integration and Continuous Delivery (CI/CD)**
   Code is integrated frequently—often multiple times per day—into a shared repository with automated builds and tests. Then, delivery pipelines automatically or semiautomatically push validated code to production, enabling more frequent, reliable releases.

3. **Automation of Repetitive Tasks**
   From build scripts and test execution to infrastructure provisioning and deployments, DevOps leans heavily on automation. This reduces manual errors and frees humans for higher-value tasks like design, optimization, and creative problem-solving.

4. **Measurement and Monitoring**
   Real-time visibility into system performance, error rates, and usage patterns is vital. Monitoring and logging solutions provide continuous feedback loops, helping teams detect issues early and guide informed decisions for improvement.

5. **Culture of Learning and Experimentation**
   DevOps encourages a blameless culture. Failures are dissected in post-mortems to glean insights and prevent repeated mistakes. Teams experiment with new tools, architectures, and improvements, iterating rapidly.

1.4 Success Stories and the Promise of DevOps

1.4.1 High-Performing Organizations

By the mid-2010s, studies from groups like **DORA (DevOps Research & Assessment)** began quantifying the performance gap between DevOps "elite performers" and traditional organizations. Elite DevOps teams deployed code **multiple times per day** (or even hundreds of times per day at large tech companies), with far fewer failures, faster recovery times, and higher job satisfaction among engineers.

> *High-performing DevOps teams deploy code **208 times more frequently** and recover from incidents **2,604 times faster** than low performers, proving that automation and collaboration drive both speed and stability.*
>
> —DORA State of DevOps Report (2024)

A few well-known examples:

- **Amazon**: Famously reached an average deployment every **11.7 seconds** at the height of its DevOps transformation, making new features instantly available and ensuring any flaws could be rolled back or patched quickly.
- **Netflix**: Developed tools like the "Simian Army" (Chaos Monkey, etc.) to automate testing of resiliency. This allowed them to deliver streaming services without significant downtime, even as infrastructure scaled exponentially.
- **Capital One**: Embarked on a DevOps journey that used cloud, CI/CD pipelines, and a cultural shift to cut release cycles in half and significantly reduce production incidents.

These examples illustrated that DevOps was **not** just for "unicorn" tech companies; traditional enterprises could adopt similar practices to achieve dramatic gains in agility and reliability.

1.4.2 Key Measurable Benefits

1. **Faster Time to Market**: Frequent releases let teams respond to business changes and user feedback more quickly.
2. **Higher Quality**: Automated testing, continuous monitoring, and immediate feedback loops help spot and fix defects sooner.

3. **Lower Risk**: Smaller, more frequent changes are easier to deploy and roll back if there's an issue, reducing the "big bang" release risk.

4. **Improved Collaboration**: Developers and operations collaborate from the start, sharing knowledge and responsibility, which fosters better relationships and fewer handoff errors.

5. **Higher Morale and Engagement**: Teams have more ownership and see their work delivered to end-users rapidly, boosting job satisfaction.

1.5 New Pressures and Emerging Challenges

Despite the substantial achievements, DevOps is not a cure-all. Many organizations encounter **stumbling blocks** on their journey:

1. **Cultural Resistance**: Surveys show that up to **45%** of DevOps initiatives stall due to cultural barriers. Middle management might resist change; ops teams can be wary of losing control to dev teams; or dev teams can fear being on-call for production incidents.

2. **Skill Gaps**: The shortage of engineers versed in DevOps and **cloud-native** technologies (Kubernetes, container orchestration, serverless, etc.) continues to limit adoption. As new tools proliferate, the learning curve steepens.

3. **Tool Sprawl and Integration**: Paradoxically, the DevOps era has seen an explosion of specialized tools for each lifecycle stage—source control, CI, security scanning, test automation, infrastructure, monitoring, etc. While specialized tools can each be best-in-class, they're often not well integrated. This can lead to "**tool sprawl**," ironically creating new silos, especially in large enterprises.

4. **Legacy Systems**: Many organizations still rely on monolithic architectures or on-premises systems built decades ago. Retro-fitting them into a DevOps pipeline can be extremely complex, requiring a major redesign or multiyear migration.

5. **Security and Compliance**: "Shift-left" security is crucial—embedding security checks into the pipeline from day one—but integrating it seamlessly remains a challenge. Strict compliance requirements (HIPAA, PCI-DSS, GDPR, etc.) can complicate automation steps. Some teams struggle to keep pace with vulnerabilities if they release code daily.

6. **Measuring and Proving Value**: DevOps success is often measured via key metrics (deployment frequency, lead time, mean time to recovery). But not all organizations track them consistently, making it difficult to prove ROI or identify where to improve.

These hurdles highlight that **DevOps is a journey**, not a one-time transformation. As software demands continue to rise—driven by user expectations of seamless, 24/7 services—teams struggle to push DevOps beyond the boundaries of its initial successes.

1.6 Toward an Expanded Vision: DevSecOps, DataOps, and NoOps

1.6.1 From DevOps to DevSecOps

Given the importance of security and compliance, many now use the term **DevSecOps** to emphasize **security** as a first-class citizen in the DevOps workflow. Code scanning, vulnerability checks, configuration audits, and threat modeling become continuous processes in the pipeline—rather than manual tasks at the end. DevSecOps ensures that security is everyone's job, from developer to operator to security engineer.

This shift is driven by

- **High-profile data breaches** highlighting the cost of insecure deployments
- **Regulatory pressure** requiring stricter audit trails, logging, and vulnerability management
- **Automation capabilities** that can embed security scanning at every commit or build

1.6.2 DataOps, MLOps, etc.

As organizations have realized data as a primary asset, new frameworks like **DataOps** have emerged. DataOps borrows from DevOps principles—continuous delivery, collaboration, and automation—to manage data pipelines and analytics processes. Similarly, **MLOps** extends DevOps to machine learning models, ensuring models are deployed rapidly yet safely, monitored for performance drift, and updated as needed.

These expansions indicate that DevOps is not just about code and servers but about **any valuable digital resource** that needs frequent, reliable, and automated updates.

1.6.3 The Rise of "NoOps"

Alongside these expansions, a bold concept took shape: **NoOps** ("No Operations"). The idea: what if infrastructure management became so automated that developers never have to think about servers, load balancers, or patching? Instead, everything is abstracted away by serverless platforms or fully managed services. In a NoOps scenario, "operations" is invisible—handled by code, automation, and intelligent systems.

Critics argue NoOps is a misnomer ("somebody, somewhere, is doing ops!"), yet the spirit of NoOps resonates. As cloud providers and container orchestration platforms become more sophisticated, the day-to-day manual tasks of provisioning, scaling, and monitoring can be heavily automated or outsourced to specialized platforms. Development teams become product-focused, iterating on features rather than wrangling servers. Still, for many organizations, NoOps remains aspirational: achieving it requires a high level of standardization, a modern cloud-native architecture, and robust automation for the entire lifecycle—plus the next frontier: **autonomous AI agents**.

1.7 DevOps Meets AI: A Glimpse Ahead

With the rise of artificial intelligence and machine learning, the **DevOps toolchain** itself is evolving. Large language models (LLMs) and specialized ML algorithms can help with

- **Predictive Analysis**: Spotting potential failures in CI/CD pipelines before they happen, suggesting fixes or improvements
- **Intelligent Monitoring and Incident Response**: Automated AIOps platforms that reduce alert fatigue by correlating logs, anomalies, and telemetry into a single root cause

- **Generative Code and Tests:** Tools like GitHub Copilot that generate boilerplate or test suites, letting developers focus on higher-level design and logic

*DevOps has already transformed software delivery, but AI promises to **redefine it again**—by shifting from automation to **autonomy**, where intelligent agents manage pipelines, tests, and deployments with minimal human intervention.*

—Google Research on AI Developer Productivity (2024)

This intersection—**DevOps + AI**—promises the next leap in productivity, setting the stage for a future where many operational tasks become autonomous, bridging us closer to a NoOps ideal. But it also raises new questions: How do we ensure data is consistent for AI to glean insights? How do we trust AI-driven suggestions or rollouts? And how do we avoid simply creating new silos in the form of half-integrated AI tools?

1.8 Change-Management Frameworks for an AI-Driven DevOps Journey

The previous sections traced DevOps from its silo-busting roots to today's AI-powered horizon. We saw how cultural resistance, skills gaps, and tool sprawl still derail transformations—even as organizations eye NoOps autonomy (see Section 1.5). What bridges that chasm is disciplined change management. The frameworks in this subchapter give leaders a tested scaffold for guiding people, process, and technology through an AI-driven DevOps evolution.

1.8.1 Why Change Management Is Nonoptional

- **High-Velocity Disruption**: AI tools iterate far faster than legacy release cadences. Without an intentional change model, "pilot sprawl" sets in—experiments never harden into muscle memory.

- **Cultural Inertia**: Up to **45%** of DevOps initiatives stall on culture alone. AI adds new fears (job loss, "black box" risk) that amplify resistance.

- **Regulatory Scrutiny**: AI-generated code and automated deployments magnify compliance exposure; auditors expect a documented, repeatable process for every change.

1.8.2 Classic Frameworks and Their Fit for AI-DevOps

Framework	Core focus	Where it shines for AI-DevOps
Kotter 8-Step	Vision and coalition-building	Rallying execs and platform teams around an "AI Paved Road" narrative; celebrates early wins (e.g., Copilot pilot) to fuel momentum
ADKAR (Prosci)	Individual adoption (awareness → reinforcement)	Coaching engineers through tooling fear: *why* AI matters, *what's in it for them*, and continuous reinforcement via metrics dashboards
Lewin 3-Phase	Unfreeze-change-refreeze	Helpful for disruptive shifts like IDE standardization or auto-merge guardrails—unsticks legacy habits, then locks new ones with policy as code
McKinsey 7-S	Org alignment (strategy, structure, skills, etc.)	Ensures AI-DevOps isn't just tooling; aligns incentives, skills matrices, and shared values across Dev, QA, Sec, Ops

> **Tip** No single model is perfect. Most high performers blend Kotter's storytelling, ADKAR's individual focus, and 7-S organizational alignment into a pragmatic "hybrid."

1.8.3 A Hybrid Playbook—A-DAIR for AI-DevOps

We propose **A-DAIR**—an adaptation of ADKAR tuned for AI:

1. **Awareness**: Share a compelling vision: *less toil, faster releases, safer code.* Use live demos of Copilot or self-healing tests to make it tangible.

2. **Desire**: Link AI benefits to personal pain points (e.g., deleting boilerplate, 30% pipeline speed-ups). Spotlight early adopters.

3. **Alignment**: Map roles, KPIs, and policy guardrails. Example: IDE standard pack + mandatory ai_source tags in telemetry.

4. **Iterate**: Roll out in sprints: pilot squad ➤ platform team ➤ org-wide. Measure DORA + AI-specific metrics in Opsera.

5. **Reinforce**: Gamify adoption (leaderboards), hold blameless AI post-mortems, and refresh prompts/policies quarterly.

1.8.4 Embedding Change Management in the DevOps Loop

DevOps phase	Change-management focus	AI-driven example
Plan	Share vision, build coalition	"AI Guild" defines prompt library and guardrails.
Code	Upskill, pair with AI	Copilot workshops; IDE extension pack enforced.
Build/test	Early wins, celebrate	Self-healing tests cut failures → showcase in town hall.
Release/deploy	Policy gates, trust	Policy-broker labels (ai-green/amber/red) guide autonomy.
Operate/monitor	Reinforce via metrics	Opsera dashboards track AI LOC, MTTR, drift patches.

1.8.5 Quick-Start Checklist

- **Nominate a Cross-Functional *AI Guild***: Include Dev, QA, Sec, Ops

- **Baseline Culture and Metrics**: Survey tool pain and capture DORA stats

- **Select a Starter Framework**: Kotter for exec storytelling + ADKAR for team adoption

- **Run a *Lighthouse* Pilot**: One service, full AI stack, and metrics in Opsera

- **Iterate and Broadcast Wins**: Internal blog posts, demo days, and CFO cost-saving reports

1.8.6 Key Takeaways

- Change management is the *engine room* that converts AI hype into lasting DevOps practice.

- Blending Kotter (vision), ADKAR (people), and McKinsey 7-S (org fitness) gives the range needed for cultural, technical, and compliance hurdles.

- Use data—lead-time, AI-accepted LOC, and drift-patch count—to reinforce behavior and silence sceptics.

- Start small, learn fast, scale deliberately. AI unlocks exponential gains only when people, process, and tech advance together.

1.9 Chapter Summary and Looking Ahead

In this chapter, we've traced the **origins and evolution** of DevOps:

1. **Siloed Beginnings**: Traditional dev and ops teams worked at odds, resulting in slow release cycles and frequent friction.

2. **Emergence of DevOps**: Inspired by Agile ideas and spurred by early adopters like Flickr, Netflix, and Amazon, DevOps became a cultural and technical movement that shortened feedback loops and improved collaboration.

3. **Core Principles**: Collaboration, automation, continuous delivery, and measurement define DevOps at its heart.

4. **Proof of Success**: Elite organizations demonstrate frequent deployments and faster incident resolution—gaining market advantage.

5. **Challenges**: Cultural resistance, skill gaps, and tool sprawl remain major roadblocks, especially in large or regulated enterprises.

6. **Toward NoOps**: As organizations look for deeper automation and consider serverless and AI, the lines between dev, ops, and security blur further, heralding a future where infrastructure "just works."

Where do we go from here? The next chapters will dive deeper into the challenges of today's fragmented DevOps ecosystems—particularly the **tool overload** and **data silos** that hamper collaboration and hamper advanced AI-driven automation. We'll then explore **the critical importance of standardization** and integrated architectures as the foundation for leveraging AI in coding, testing, infrastructure, and release orchestration. Finally, we'll see how multiagent AI systems can push DevOps closer to a NoOps reality—where the pipeline practically runs itself and humans focus on innovation rather than firefighting.

As you continue reading, keep in mind that DevOps is less a destination and more a **continuous journey**. The journey is about **aligning people, process, and technology** so that software—and by extension, the business—can evolve at the speed of customer demand. AI promises to accelerate this evolution dramatically, but it depends on a stable base of standardized, integrated tooling and data. That's the next chapter's focal point: understanding the **fragmented state** of DevOps today and why it's so urgent to unify and standardize before layering AI on top.

1.10 Key Takeaways

1. **DevOps Origin**

   - Evolved as a response to siloed dev and ops teams, inspired by Agile principles.
   - Early pioneers (Patrick Debois, John Allspaw, Paul Hammond) showcased how frequent, reliable deployments could be done at scale.

2. **Cultural and Technical Movement**

   - Emphasizes collaboration, continuous integration, continuous delivery, and measurement.
   - Automation is central: from builds to testing to deployments.

3. **Proven Impact**

   - Organizations like Amazon, Netflix, and Capital One exemplify how DevOps can accelerate releases while reducing errors.
   - Studies show DevOps correlates with higher quality software and happier teams.

4. **Challenges Remain**

   - Cultural resistance, legacy systems, tool sprawl, and security integration slow adoption.
   - Skill shortages and organizational inertia are common impediments.

CHAPTER 1   THE EVOLUTION OF DEVOPS

5. **From DevOps to NoOps**

    - The desire for ever-greater automation leads some to envision "NoOps," where infrastructure concerns disappear behind fully managed or serverless platforms.

    - AI is emerging as a key enabler, promising to handle operational tasks autonomously—if data and processes are standardized enough to support it.

With a historical perspective in place, we now turn to a pressing question: **If DevOps is so effective, why do so many teams still struggle?** Chapter 2 dives into the **fragmentation problem**—how multiple, disjointed tools and siloed data hamper the potential of DevOps, leading us to see why standardization is not just a buzzword but an essential stepping stone to an AI-empowered, NoOps future.

CHAPTER 2

Fragmented Software Development: Why DevOps Isn't Always Enough

Despite the proven benefits of DevOps, many organizations find themselves stuck. They've adopted continuous integration (CI), continuous delivery (CD), and cloud infrastructure—but the everyday reality is anything but seamless. In practice, **tool sprawl**, **siloed data**, and **disconnected teams** can derail even the best DevOps intentions. This chapter dives into the heart of that fragmentation: how a patchwork of specialized tools can create new silos, how data gets scattered across systems, and how these disconnects erode collaboration. Understanding these challenges is the first step to implementing the **standardized, integrated architectures** needed to fully leverage AI and move closer to a NoOps future.

CHAPTER 2   FRAGMENTED SOFTWARE DEVELOPMENT: WHY DEVOPS ISN'T ALWAYS ENOUGH

2.1 The Rise of Tool Sprawl

2.1.1 The Allure of Specialized Tools

Ironically, DevOps—meant to streamline software delivery—has spawned a **massive ecosystem of niche tools**. There's a tool for everything: source control, CI servers, test automation, release orchestration, configuration management, container orchestration, security scanning, monitoring, log aggregation, analytics, and more. Each solves a specific pain point exceptionally well. For instance:

- **Git** (or GitHub, GitLab, Bitbucket) for version control
- **Jenkins**, **CircleCI**, or **Bamboo** for continuous integration
- **SonarQube** or **Snyk** for static code analysis and vulnerability scanning
- **Terraform**, **CloudFormation**, or **Pulumi** for infrastructure as code (IaC)
- **Splunk**, **Datadog**, or **New Relic** for monitoring and observability

At first glance, adopting the "best tool for the job" in each category seems like a no-brainer. However, what often starts as a logical approach can balloon into a **patchwork** of 20, 30, or even more discrete tools and platforms—each with its own interface, usage model, data format, and integration points.

2.1.2 Number of Tools and the "Tool Tax"

*Over **50% of enterprises** report using more than **20 DevOps tools**, leading to tool sprawl that increases **complexity, cognitive load, and maintenance overhead**, rather than improving productivity.*

—2016 DevOps Toolchain Survey

Maintaining such a broad toolset incurs what's often called a "**tool tax**":

- **Licensing or subscription costs** for each product, which can become significant if usage scales widely across teams.

- **Integration Overhead**: Hooking up each tool to the next (e.g., plugging your CI system into your source control, your security scanner into your CI, your test results into your reporting dashboard, etc.).

- **Context Switching**: Developers and operators might bounce between multiple interfaces (like Jenkins for builds, JIRA for tickets, Slack for notifications, Splunk for logs, etc.). Each switch demands reorientation, slowing the team.

- **Support and Training**: Every additional tool means more specialized knowledge to master. As staff turnover or reorgs occur, new hires must learn an ever-growing list of systems.

This overhead might be tolerable in small doses, but as the toolchain expands, the friction grows exponentially. In many organizations, the very **DevOps** improvements (faster releases, greater automation) become undercut by the complexity of orchestrating so many separate platforms.

2.2 Data Silos and Lack of End-to-End Visibility

2.2.1 Fragmented Data Landscape

Each tool in the chain often **stores and formats data differently**. Build logs remain on one CI server, while test results live in another database, deployment records in yet another, and performance metrics in a separate monitoring system. This creates **data silos**, with each silo containing only part of the overall puzzle.

> *Without a unified toolchain, teams waste hours **reconciling disconnected logs, metrics, and test results**, making debugging and incident resolution far slower than it should be.*
>
> —Google Site Reliability Engineering (SRE) Principles

For example, consider a typical "Day 2" operations scenario:

1. Your CI tool says the latest build passed all tests.
2. Your container registry shows an image tagged `v2.1.0`.
3. Your infrastructure-as-code logs might show that version `v2.1.0` was deployed to staging.
4. Your monitoring solution indicates a spike in latency at 1:23 PM.
5. Your logging solution captures a flood of error messages from 1:24 PM to 1:26 PM.

But **tying these threads together**—so you can see exactly which code change caused the spike—is not straightforward unless you have an integrated system that can correlate build artifacts, deployment logs, and runtime metrics. In many DevOps shops, an engineer must **manually**

piece together logs from multiple systems to reconstruct the story. This process slows down root cause analysis and leads to **longer mean time to recovery (MTTR)**.

2.2.2 The Visibility Gap

Siloed data also means **lack of a single source of truth**. When issues arise—like a performance regression or a security vulnerability—there's no unified dashboard that instantly shows

- The relevant code commits and authors
- The associated build/test results
- The configuration changes or environment variables
- The application logs and user analytics around the time of incident
- The known vulnerabilities or compliance flags

Instead, each piece of data must be pulled from a separate tool. In a perfect world, your pipeline would unify this data under an automated "Software Bill of Materials" or "Chain of Custody" concept. But in practice, **fragmentation** is the norm. Research indicates that **74%** of DevOps teams **lack end-to-end visibility** across their entire toolchain. Not only does this hamper troubleshooting, but it also makes it tough to measure success metrics like lead time or deployment frequency.

2.3 Impact on Collaboration and Workflow

2.3.1 DevOps Irony: New Silos

DevOps was meant to erase silos between dev and ops. Yet, ironically, many organizations now suffer from **tool-based silos**. Different teams—say QA, security, release engineering, or business analysts—may cling to their own specialized tools and processes. As a result:

- **QA teams** might primarily use one test management platform, rarely checking the pipeline's integrated view.

- **Security** might run separate vulnerability scans or penetration tests, with results stored in a standalone system.

- **Ops** might track production changes in a platform that dev teams rarely see.

This partial overlap fosters miscommunication. Teams speak the same "DevOps language" but operate in **separate digital ecosystems**. Paradoxically, **the more specialized the tools, the harder it can be to unify** them. The outcome is the very fragmentation DevOps sought to cure, only now it's scattered across multiple SaaS or on-prem solutions.

2.3.2 Collaboration Friction and Context Switching

When each department or team uses different platforms, **collaboration friction** arises:

- **Context Switching**: A developer investigating a production issue might need to bounce between the APM tool, the deployment logs in a separate console, and Slack messages with ops—each requiring time to open, authenticate, search, and correlate.

- **Duplicate Efforts**: Teams often duplicate data entry in multiple systems (e.g., logging defects in JIRA but also having to reference them in a separate QA tool).

- **Misaligned Ownership:** If an incident arises at the boundary between two specialized tools, no one is sure which group is responsible. Delays mount as teams debate who "owns" the fix.

A hallmark of DevOps success is **fast feedback loops**. Tool sprawl can defeat this by inserting friction at every handoff, undermining the speed and agility DevOps is supposed to deliver.

2.4 The Culture of "Choose Your Own Tool"

It's not all gloom—allowing teams the freedom to pick tools can **spark innovation**. Many DevOps success stories do involve a "grassroots" approach where each team quickly adopts the best tool for its function. The trouble arises when **no overarching strategy** or governance emerges. Over time, you end up with

- Multiple code repositories (GitHub, Bitbucket, GitLab) each storing separate pieces of the same product
- Multiple CI systems—maybe TeamCity for some groups, Jenkins for another, and GitHub Actions for the rest
- Inconsistent practices—some teams do canary releases, others do blue-green, others do big bang deployments

Soon, you can't easily share pipeline templates or best practices because the environment differs drastically across teams. The overhead grows. Meanwhile, new hires struggle to figure out which tools they need for which project. Over time, what started as flexible autonomy morphs into **chaotic fragmentation**.

2.5 The Hidden Costs of Fragmentation

2.5.1 Slowed Time to Market

One might assume more tools = faster releases. However, when those tools aren't integrated, the **cumulative friction** actually slows things down. Repetitive tasks—like re-authenticating, copying artifacts between systems, manually updating statuses—waste valuable cycles. Delays can compound as tasks wait for the "right" person who has the knowledge to navigate a certain tool.

2.5.2 Increased Risk of Errors

Manual data transfer or reconfiguration often leads to human error. For instance, an environment variable set in one pipeline might not propagate to another if the systems don't sync. Similarly, the QA environment might unknowingly be on an older build because the test orchestration tool wasn't updated, while the production environment is on a newer, untested build.

Security vulnerabilities also slip through cracks if scans or patch processes are inconsistent. If the security tool is disconnected from CI/CD, it may not catch newly introduced vulnerabilities. Fragmentation can open the door for compliance violations if no one is truly sure which version of the software is running where.

2.5.3 Lower Morale and Higher Burnout

Developers and operators typically crave efficiency. Wasting hours on searching for the right logs or toggling between multiple dashboards can be demoralizing. The **cognitive load** alone can contribute to stress and burnout. In a domain (DevOps) that already includes on-call duty and complex system design, the added friction from scattered tools can push engineers toward frustration—or turnover.

2.5.4 Difficulty Scaling

As organizations grow, the issues multiply. What worked for 2–3 teams breaks down when you have 20–30 teams. The best DevOps transformations rely on a consistent and repeatable pipeline model. With fragmentation, it's **impossible** to scale consistent processes. Each new team might adopt yet another specialized tool, further compounding the sprawl.

2.6 Real-World Example: A Financial Services Firm in "Tool Chaos"

2.6.1 Multiple CI/CD Tools, Repos, and Scripts

Consider a mid-sized financial services company that jumped on the DevOps bandwagon. Initially, each development squad was empowered to choose tools for code hosting, CI, and monitoring. Squad A used GitHub and Jenkins, while Squad B tried Bitbucket and Bamboo. QA teams liked different test frameworks. Operations used custom scripts for deployments on VMs, while a separate "Cloud Ops" group began using containers with AWS ECS.

After a few years:

- **More than eight different CI pipelines** were running in parallel, each with unique scripts and plug-in versions.
- **Three distinct code repositories** (GitHub, Bitbucket, an internal Git server) existed.

- Logging was split among **Splunk, Elasticsearch**, and a legacy solution.

- Automation scripts were written in **Bash, Python, and some in PowerShell**—all doing roughly the same tasks.

- Security scans happened sporadically, triggered by an external team that rarely integrated with the squads' pipelines.

Developers had to learn multiple dashboards to troubleshoot. When a release caused a production issue, engineers spent hours reconciling conflicting logs just to confirm which microservice version was even deployed. Meanwhile, management complained: "We invested in DevOps, so why are we still so slow?"

Ultimately, this company formed a **platform engineering** team to standardize around a single set of repositories, pipelines, and integrated logging and monitoring—reducing duplication and accelerating new feature rollouts. This shift took months of effort but was essential to break the fragmentation.

2.6.2 The Complexity of Multiple IDEs

Further complicating matters, each team used a different **IDE (Integrated Development Environment)** or code editor:

- Some developers preferred **Visual Studio** (or VS Code) for .NET or JavaScript.

- Others used **IntelliJ** or **PyCharm** from the JetBrains suite for Java or Python.

- A few front-end teams stuck to **Atom** or **Sublime Text**, citing faster startup or personal preference.

- Still others used **Eclipse** due to legacy plug-ins.

CHAPTER 2   FRAGMENTED SOFTWARE DEVELOPMENT: WHY DEVOPS ISN'T ALWAYS ENOUGH

While autonomy in editor choice can boost individual productivity or comfort, it also **fractured the developer experience**. Plug-ins to integrate with the CI system or code scanning had to be re-implemented for each editor. Consistent coding standards were harder to enforce across so many IDE ecosystems. In some cases, code-level collaboration (like real-time pair programming or shared debugging) became cumbersome because half the team used a JetBrains product and the other half was on Visual Studio.

Worse, the editors weren't centrally integrated with the rest of the pipeline—meaning developers would manually switch to a separate web console to check build results or run deployment scripts. This constant **context switching** killed the "flow" state that fosters deep productivity. In turn, bugs slipped through, and knowledge of best practices was often locked in siloed editor configurations.

Toward a Standard Editor and Cloud Workspace As part of the platform engineering shift, the company began piloting a **standardized editor**—in this case, Visual Studio Code plus a curated set of extensions for Docker, Kubernetes, linting, and integrated CI/CD tasks. Their long-term plan was to migrate code editing **fully to the cloud**, so that every developer session could be container-based, ephemeral, and automatically linked to the company's Git repos and pipelines. Although some veteran developers were initially resistant, the promise of consistent dev environments, AI-assisted coding features, and frictionless collaboration made many see the potential for a **more unified developer experience**.

This story underscores that **fragmentation goes beyond DevOps toolchains**: it can extend right into each developer's local environment. Inconsistent IDEs and scattered plug-ins create friction, hamper code quality, and impede the broader DevOps vision. By standardizing on a single or at least a well-integrated family of editors—and eventually moving to **cloud-based IDEs**—teams can reduce configuration drift, streamline onboarding, and stay "in the flow" more consistently. It also lays the groundwork for future AI-driven coding and testing capabilities (covered in later chapters).

2.7 The AI Readiness Angle

Why does **tool sprawl** matter so much when we talk about AI? Because **AI thrives on data**—the more consistent, complete, and high-quality your data, the better AI models can analyze it and produce meaningful insights. If logs, metrics, builds, and test results are all scattered, any attempt at using AI for anomaly detection, predictive scaling, or automated code generation will face blind spots.

Moreover, AI-based DevOps solutions—like "intelligent test selection" or "auto-remediation scripts"—need **a unified view** of the entire software lifecycle to make good decisions. A half-baked integration that only sees partial data can create erroneous or even harmful outputs (e.g., rolling back the wrong service). In short, **fragmentation is the enemy** of advanced AI-driven DevOps.

2.8 Why Fragmentation Persists

Despite the obvious drawbacks, fragmentation endures because

1. **Individual Teams Optimize Locally:** Each group chooses the best immediate solution, rather than adopting a standardized approach.

2. **Organic Growth:** Tools accumulate organically as new services are spun up, M&A occurs, or new leaders bring their favorite solutions.

3. **Lack of Executive Mandate:** Without strong leadership pushing for an integrated platform or "golden pipeline," the default is tool chaos.

4. **Short-Term Gains, Long-Term Costs:** Each tool might bring a short-term productivity boost, but the cumulative overhead over time gets overlooked.

5. **Fear of Breaking Existing Workflows**: Teams worry that standardizing or migrating to a single platform will cause disruptions or require steep learning curves.

2.9 The Way Forward

2.9.1 Recognize the Cost

The first step in addressing fragmentation is acknowledging it as a **serious business cost**—not just a mild inconvenience. When leadership sees the impact on time-to-market, software quality, incident response, and staff morale, they are more likely to support a unification effort.

2.9.2 Plan for Standardization

Chapter 3 will dive deeper into **why standardization is crucial** and how to achieve it without stifling innovation. From adopting a standardized "platform" to establishing consistent processes (e.g., branching strategies, pipeline templates, environment naming), there are clear steps to ensure that DevOps becomes truly integrated, not a labyrinth of specialized tools.

2.9.3 Evolve from DevOps to Platform Engineering

Many companies are forming **platform engineering** teams whose job is to provide **self-service, integrated pipelines** that unify tools while still letting teams pick specialized solutions if they adhere to a consistent interface. This approach balances standardization with flexibility and sets the foundation for advanced AI adoption.

2.9.4 Focus on Data Centralization

A consolidated data lake or "single pane of glass" for DevOps metrics, logs, and artifacts is key. By streaming build logs, test results, performance metrics, and security scans into a **common data layer**, organizations can break down silos and **enable AI** to deliver real insights. Tools that unify data (e.g., a robust platform that automatically tracks commits, deployments, and runtime metrics) reduce the manual overhead of correlation.

2.9.5 Standardize the Developer Experience

Beyond unifying CI/CD and monitoring, **choose a consistent, well-integrated IDE or set of editors** that ties directly into the pipeline and testing frameworks. This can involve

- A minimal range of **officially supported editors** (e.g., Visual Studio Code and IntelliJ) with curated extensions

- Plans to **move IDEs to the cloud**, offering ephemeral, containerized dev environments preconfigured with dev/test tools

- Enforcing consistent linting, code formatting, and code review workflows across all teams, so developers stay "in the flow" rather than juggling multiple local setups

This approach not only reduces friction but also **paves the way** for AI-driven coding assistants—since those assistants can hook into a single, standardized environment to generate or refactor code seamlessly.

2.10 Chapter Summary and Looking Ahead

In this chapter, we've examined the **fragmentation plague** that persists in many DevOps environments:

- **Tool Sprawl**: Too many specialized tools lacking robust integration.

- **Data Silos**: Disconnected logs, metrics, and build artifacts hamper end-to-end visibility.

- **Collaboration Friction**: New silos pop up around specialized solutions, ironically defeating DevOps' collaboration goals.

- **High Hidden Costs**: Slower deliveries, increased risk of misconfigurations, lower morale, and difficulty scaling.

- **AI Roadblock**: Fragmented data and inconsistent editor setups undermine AI's potential.

We also **introduced** the **complexity of multiple IDEs**, which can scatter developer experiences and hamper efforts to implement consistent security checks, code quality standards, and advanced AI assistants.

For DevOps to evolve toward a **truly automated**, AI-empowered, and eventually NoOps future, teams must address these fragmentation woes. **Standardization** is not just a buzzword; it's the key to unlocking the next stage of DevOps maturity—both in toolchains **and** in the developer experience. Chapter 3 will show why focusing on standardizing toolsets, data, and processes is essential—and how organizations can do it **without** stifling innovation or imposing rigid mandates. By unifying the core pipeline and developer environment, we create the "data infrastructure" and consistent dev flows that AI needs to thrive, propelling DevOps to its next frontier.

2.11 Key Takeaways

1. **Tool Proliferation**
   - While picking the "best tool for the job" can be beneficial, it often leads to fragmentation and high overhead ("tool tax").
   - Over 50% of large orgs use 20+ DevOps tools, creating complexity and slowing delivery.

2. **Data Silos Harm Visibility**
   - Logs, metrics, builds, and test results scattered across multiple platforms make it difficult to pinpoint root causes or track deployments accurately.
   - Over 70% of teams lack true end-to-end visibility in their pipelines, extending incident response times.

3. **Collaboration Friction**
   - Instead of bridging dev and ops, poorly integrated toolchains can create new silos around specialized solutions.
   - Context switching and duplicated effort reduce productivity and morale.

4. **Multiple IDEs Complicate Developer Flow**
   - When each team adopts a different code editor or IDE, consistency in code quality, security checks, and plug-in support suffers.
   - Future AI-driven coding/testing solutions rely on standardized, integrated developer environments for seamless integration.

5. **Hidden Costs of Fragmentation**
   - Slowed release velocity, increased risk of errors, and higher burnout.
   - Potential security holes if scans or compliance checks aren't consistently integrated.

6. **AI Readiness**
   - AI-driven DevOps requires **unified, high-quality data** to provide accurate insights. Fragmentation starves AI of the correlated data it needs.
   - An inconsistent developer experience also hinders AI coding assistants from scaling effectively.

7. **Path Forward**
   - Acknowledge the costs of fragmentation, plan for **platform engineering**, and centralize data and the development environment.
   - Standardizing IDEs and eventually moving them **to the cloud** can reduce friction, unlock AI integrations, and keep developers in flow.

With the fragmentation issue laid bare, we're ready to explore the crucial topic of **standardization** in the next chapter—both at the pipeline level and within the developer experience. By unifying these elements, teams can finally realize the true potential of DevOps, setting the foundation for advanced AI capabilities and marching steadily toward the NoOps horizon.

CHAPTER 3

The Case for Standardization: Building the Foundation for NoOps

In the previous chapters, we explored how DevOps arose to break down silos and speed up software delivery—only to discover that **fragmented toolchains** and **heterogeneous developer environments** can undermine its effectiveness. From sprawling CI/CD pipelines to multiple, unintegrated IDEs, fragmentation not only slows teams but also blocks the potential of AI-driven automation. In this chapter, we make the **case for standardization**: establishing consistent processes, toolchains, and developer experiences. We'll see how standardization *enables* innovation—rather than stifling it—and why it is **essential** for AI readiness, compliance, and, ultimately, the realization of a NoOps future.

CHAPTER 3 THE CASE FOR STANDARDIZATION: BUILDING THE FOUNDATION FOR NOOPS

3.1 What Do We Mean by "Standardization"?

3.1.1 Defining Standardization in DevOps

Standardization often evokes images of rigid bureaucracy or one-size-fits-all mandates, but in a DevOps context, it's about creating **a unified, repeatable, and data-friendly framework** for the entire software delivery lifecycle. Concretely, standardization might include

- **A "Golden Pipeline" Approach**: Using a common continuous integration/continuous delivery (CI/CD) template or platform across teams, so everyone follows consistent build, test, and deployment steps

- **A Curated Set of Tools**: Limiting the proliferation of overlapping or redundant solutions in source control, test automation, monitoring, or security scanning

- **Consistent Developer Environments**: Adopting a single or minimal set of IDEs, coding standards, and code review processes—often with preconfigured plug-ins or extensions to unify the experience

- **Unified Data Flows**: Centralizing logs, metrics, build artifacts, and test results so they can be easily correlated and analyzed, especially by AI or analytics systems

In short, standardization **reduces variability** in how software is built, tested, deployed, and monitored. Done right, it preserves enough flexibility for teams to adapt to unique needs while still ensuring that critical pieces—like security checks or code-quality gates—aren't optional.

3.1.2 Why It Matters More Than Ever

As organizations scale their DevOps efforts, the **complexity of multiple squads, microservices, and cross-functional workflows** magnifies the need for standardization. Without shared guardrails or common frameworks, each team's local optimizations accumulate into system-wide chaos. Meanwhile, advanced initiatives—like AI-driven testing or self-healing infrastructure—require **clean, consistent data** from the entire pipeline, which only emerges when the pipeline is well integrated and standardized.

3.2 The Core Benefits of Standardization

3.2.1 Streamlined Collaboration

When each team uses **the same fundamental toolchain** or at least shares a consistent set of integrations and naming conventions, collaboration becomes much simpler:

- **Less Context Switching**: A developer from Team A can quickly move to Team B's repo or pipeline without learning an entirely new interface or script language.

- **Unified Documentation**: Instead of referencing half a dozen "How to deploy" guides, you have a single or minimal set of docs describing how to run builds, tests, or rollbacks.

- **Shared Language**: Everyone can talk about "the pipeline" or "the environment" with the same assumptions, building a stronger DevOps culture.

3.2.2 Reduced Operational Overhead (the Anti-Tool-Tax)

A standardized toolkit cuts down on

- **Licensing Sprawl**: Minimizing the number of overlapping solutions for the same function.
- **Maintenance**: Fewer upgrade paths, fewer integration breakpoints when moving from one version of a tool to another.
- **Training and Onboarding**: New hires ramp up faster when they only need to learn one or two platforms, not ten.

Moreover, platform or DevOps engineers can focus on **deep expertise** in a smaller set of tools, improving the overall reliability of the pipeline.

3.2.3 Stronger Security and Compliance

By standardizing on

- **Approved toolchains** with built-in security checks
- **Consistent pipeline templates** that embed scanning, gating, and auditing
- **Unified environment provisioning** (e.g., infrastructure as code with the same Terraform modules or Helm charts)

you ensure security and compliance are **applied uniformly** across teams. This is essential for meeting regulatory demands (HIPAA, PCI-DSS, GDPR, etc.). Auditors or security teams can focus on verifying **one pipeline** rather than countless custom-coded release scripts.

CHAPTER 3   THE CASE FOR STANDARDIZATION: BUILDING THE FOUNDATION FOR NOOPS

3.2.4 Increased AI Readiness

AI thrives on large, consistent datasets and the ability to correlate them. If every pipeline, environment, or editor plug-in produces data in a different structure, feeding it into an AI model becomes a monumental headache. Standardization ensures

- **Uniform build/test logs** that can be parsed and tagged automatically
- **Consistent metadata** about commits, artifacts, security scans, and runtime metrics
- **Predictable structure** that advanced automation tools (like AI-based test generation or anomaly detection) can rely on

With standardized pipelines and developer environments, it's far more feasible to introduce **autonomous agents** that handle tasks such as predictive scaling, risk-based testing, or even auto-remediating code vulnerabilities.

3.3 Addressing Fears and Misconceptions

3.3.1 "Won't Standardization Kill Innovation?"

A common pushback is that standardizing the pipeline or IDE usage will hamper creativity. In reality, **innovation in DevOps** often **increases** once teams offload the complexity of basic scaffolding. Developers can still innovate on features, architectures, or test strategies—they just do so **within** a stable, automated environment that handles the repetitive details. Moreover, organizations can allow "opt-out" or "innovation tracks" for carefully vetted exceptions. The key is to keep standardization from becoming a straitjacket by periodically reviewing and updating the "blessed" tools or processes to incorporate emerging technologies that prove genuinely beneficial.

3.3.2 "It's Too Hard to Switch from Existing Tools"

Teams worry about migrating away from their pet solutions. But often, the cost of continuing fragmentation is higher. Phased migrations—starting with **new projects or new teams** adopting the standardized pipeline—can mitigate disruption. Over time, older projects can either be retired or refactored to the new approach. Some organizations run "**center of excellence**" or "**lighthouse**" projects to demonstrate the value of standardization at a small scale, then roll out more broadly.

3.3.3 "We Need Different Tools for Different Languages or Frameworks"

In large companies, it's true that a single CI engine or code editor might not suit all languages. But that doesn't preclude standardization. Many modern platforms (e.g., Jenkins, GitHub Actions) support a wide variety of languages via plug-ins. Similarly, an editor like Visual Studio Code or IntelliJ can handle multiple languages with the right extensions. The goal is not forced homogeneity but a **manageable** variety—perhaps allowing two or three CI solutions or IDEs max, each thoroughly integrated, rather than a dozen uncoordinated stacks.

3.4 Approaches to Standardization

3.4.1 Platform Engineering and the Internal Developer Platform

An increasingly popular strategy is forming a **platform engineering** team responsible for building and maintaining an **internal developer platform**. This platform includes

- **A self-service portal** where developers can provision standard pipelines, environments, and code repos with a few clicks
- **Prebuilt CI/CD templates** that automatically embed security scans, test automation, and environment provisioning
- **Shared services** like container registries, artifact storage, monitoring dashboards, and compliance checks

By offering these capabilities via a central platform, you incentivize teams to use the standard approach—**because it's the easiest path**. The platform can be flexible enough to let squads choose certain stack details (e.g., Node.js vs. Python) while still enforcing consistent DevOps practices.

3.4.2 Reference Architectures and Golden Pipelines

Publishing **reference architectures**—complete with sample code, pipeline configurations, and recommended tool integrations—helps teams adopt best practices rapidly. A "golden pipeline" might define, for example:

1. **How code is branched** (e.g., trunk-based vs. Gitflow)
2. **Which tests run on commit** vs. nightly vs. prerelease

3. **Automated security steps** (static analysis, dependency checks)

4. **Deployment patterns** (blue-green or canary) for staging/production

5. **Monitoring and alerting** defaults (hooks into a standard observability stack)

Each new project or microservice can clone the golden pipeline, drastically reducing guesswork and ensuring compliance with organizational standards.

3.4.3 Standardizing the Developer Experience

Chapter 2 highlighted how multiple IDEs hamper consistency. Standardizing on one or two editors, with curated extension packs (for linting, debugging, or cloud integration), ensures uniform code quality checks, consistent local builds, and a smoother on-ramp to future **cloud-based IDEs:**

- **Visual Studio Code** or **IntelliJ** as the baseline, configured to automatically load environment variables, authenticate with the chosen SCM, and run local tests exactly as the CI pipeline would

- **Optional**: Ephemeral dev containers in the cloud, so developers can spin up a preconfigured environment that includes the latest build tools and security checks, all managed by the platform engineering team

CHAPTER 3   THE CASE FOR STANDARDIZATION: BUILDING THE FOUNDATION FOR NOOPS

3.4.4 Data Unification

Possibly the most critical aspect is ensuring **all logs, metrics, build artifacts, test results, and security scans** feed into a central repository or data lake. Whether you use an ELK stack, Datadog, Splunk, or a custom data warehouse, the point is to unify everything. This means normalizing log formats (e.g., JSON), tagging data with consistent metadata (commit ID, service name, environment), and building an easy interface (or API) to query the entire pipeline's history.

3.5 Standardization As the Launchpad for AI

3.5.1 AI Demands High-Quality Data

An inconsistent, siloed environment starves AI-based systems of the correlated, comprehensive data they need. On the other hand, a standardized pipeline that tags every artifact and logs every event with meaningful metadata becomes a **treasure trove** for AI solutions. This is especially true if your organization wants to implement

- **Intelligent Test Selection**: An AI that decides which subset of tests to run based on code changes, requiring historical data on test coverage, commits, and code complexity.

- **Predictive Analytics**: Using metrics from prior releases to predict production incidents or performance regressions.

- **Generative Code and Fixes**: From AI pair programming to automated security fixes, the AI needs consistent references to code style guidelines, libraries, and environment configs.

CHAPTER 3   THE CASE FOR STANDARDIZATION: BUILDING THE FOUNDATION FOR NOOPS

*AI thrives on structured, consistent data—without standardization in **toolchains, workflows, and metadata**, AI-driven automation will never reach its full potential.*

—Digital.ai DevOps Governance Reports

3.5.2 Enabling Autonomous Agents

Later chapters will discuss **autonomous, multiagent systems** that can orchestrate dev, test, and ops tasks. For these agents to function effectively—rolling back a faulty release, auto-scaling the environment, or patching a known vulnerability—they need a single, standardized "view" of the pipeline. If half the code is in GitHub but the other half is in a local Git server, or if half the logs are missing crucial metadata, the AI cannot reliably act. In that sense, standardization is the **foundation** that NoOps automation stands upon.

3.6 Case Study: A Global Tech Firm's "Platform First" Approach

Take the example of a global software company with 50+ microservices across multiple business units. Initially, each unit had its own CI/CD, usage of Git repos, and monitoring solutions. Post a major outage triggered by an unnoticed environment mismatch, leadership mandated a **"Platform First"** approach:

1. **Platform Engineering Team:** They established a cross-functional group of senior engineers responsible for building a single, integrated developer platform.

CHAPTER 3   THE CASE FOR STANDARDIZATION: BUILDING THE FOUNDATION FOR NOOPS

2. **Unified SCM and Pipelines**: They standardized on GitHub and GitHub Actions, designing workflows that included mandatory linting, code scanning, and automated test suites.

3. **IDE Standardization**: They offered a curated set of Visual Studio Code extensions that matched the pipeline's capabilities (e.g., Docker, Kubernetes, Terraform). A small JetBrains alternative was supported for teams with specialized needs.

4. **Data Consolidation**: All logs, from dev to production, were routed into an ELK stack, enriched with consistent tags for service name, environment, and version.

5. **Migration Roadmap**: They allowed each business unit a six-month window to adopt the new platform. Any new project had to start on it by default.

6. **Results**: Within a year, the firm saw a **40% reduction** in average lead time for changes. Incident resolution improved dramatically due to cross-service visibility. They are now exploring AI ops tools to predict capacity needs and to detect anomalies in logs. By centralizing data and processes, these advanced AI capabilities are far easier to integrate.

This case illustrates how standardization needn't be "draconian." When done thoughtfully, it can catalyze efficiency, reduce firefighting, and clear the path for next-level AI-driven DevOps.

3.7 Chapter Summary and Looking Ahead

Standardization may sound counterintuitive in a culture that prizes speed and autonomy, but **chaotic fragmentation** ultimately undermines DevOps goals. By adopting consistent pipelines, toolchains, developer environments, and data models, organizations set the stage for

- **Streamlined collaboration and faster onboarding**
- **Reduced risk and compliance overhead**
- **Better operational efficiency** ("anti-tool-tax")
- **AI readiness**, enabling advanced analytics, intelligent testing, and autonomous multiagent systems
- A clear **path to NoOps**, where operation tasks can become invisible or fully automated

In the next chapter, we'll dive into **cloud-native architectures** and how they synergize with standardization to create **unified, data-centric environments.** We'll explore how microservices, containers, and infrastructure as code can be integrated in a standard, cloud-based platform—laying the technical groundwork for seamless scaling and AI-driven automation.

3.8 Key Takeaways

1. **Standardization Defined**
    - Establishing a **unified, repeatable** framework for DevOps, from pipeline to IDE, fosters consistency and data integrity.

2. **Benefits**

    - **Collaboration:** Less friction and faster handoffs.
    - **Security and Compliance**: Uniform scanning and approvals.
    - **Reduced Overhead**: Less duplication and "tool tax."
    - **AI Readiness**: Clean, correlated data for advanced automation.

3. **Misconceptions**

    - Standardization need not squash innovation. Instead, it **accelerates** it by removing boilerplate tasks.
    - Migration can be phased, starting with new projects or "lighthouse" teams.

4. **Key Approaches**

    - **Platform Engineering:** A dedicated team providing self-service pipelines and shared services.
    - **Reference Architectures and Golden Pipelines**: Templates that embody best practices.
    - **IDE Unification**: Narrowing the range of editors and aligning them with the pipeline.
    - **Data Unification**: Centralizing logs, metrics, test results, etc.

5. **NoOps Enabler**
    - Standardization is **the backbone** for AI-based DevOps. Without consistent data and processes, AI cannot reliably automate tasks or unify the pipeline.
    - The future of DevOps—autonomous agents, predictive scaling, AI-driven testing—depends on a stable, standardized foundation.

Armed with a clearer view of **why** standardization matters and **how** it can be approached, we're ready to explore **cloud-native architectures**—the next puzzle piece in building an integrated environment that paves the way toward NoOps. After all, standardization alone won't solve everything unless the underlying infrastructure also embraces modern, API-driven, container-friendly practices—an arena ripe for further automation and AI.

CHAPTER 4

Cloud-Native and Data-Centric Approaches

Standardization (as discussed in Chapter 3) provides the **foundation** for consistent DevOps practices and AI readiness. However, true transformation requires **modernizing the underlying infrastructure** so your pipelines and applications can dynamically scale, adapt, and capture the data needed for continuous optimization. In this chapter, we focus on **cloud-native and data-centric architectures**—an approach that prioritizes microservices, containers, infrastructure as code (IaC), and consolidated observability. Adopting these principles paves the way for agile scaling, reliable delivery, and advanced automation, including the AI-driven NoOps paradigm we explore later.

4.1 Why "Cloud-Native" Matters

4.1.1 Definition and Core Principles

Cloud-native typically refers to designing systems specifically to leverage **cloud environments**, rather than simply lifting a traditional on-premises application into the cloud. A cloud-native architecture exhibits

- **Microservices** instead of monoliths, so each feature can be developed, deployed, and scaled independently

- **Containerization** (e.g., Docker) to encapsulate services in portable, lightweight runtime units

- **Dynamic orchestration** (like Kubernetes) to manage containers, scaling, load balancing, and failover automatically

- **API-driven communication**, so microservices talk over well-defined APIs (usually REST or gRPC)

- **Automated infrastructure** (e.g., infrastructure as code) ensuring environments can be provisioned, updated, and torn down reliably

Where a "traditional" approach might revolve around large, fixed servers, manual configuration, and occasional big bang releases, cloud-native shifts to ephemeral resources, continuous updates, and **self-healing** platforms. This synergy with DevOps fosters shorter release cycles, **instant availability** of new capabilities, and more resilient applications—**especially crucial** when AI tooling evolves at breakneck speed.

4.1.2 The Shift from Monoliths to Microservices

Early in software development, many teams built **monolithic applications**—all features in one codebase, deployed as a single package. While straightforward at first, monoliths become cumbersome as they grow:

- Minor changes require retesting or redeploying the entire monolith.

- Scaling means scaling the whole app, even if only one module needs more capacity.

CHAPTER 4 CLOUD-NATIVE AND DATA-CENTRIC APPROACHES

- Code merges get riskier, slowing down release frequency.

Moving to a **microservices** model addresses these pain points:

- Each service can have its own CI/CD pipeline and be tested, versioned, and deployed independently.
- Teams gain autonomy—one squad can focus on a "payments" service, while another owns "notifications."
- Scaling is selective; if the payments service sees high traffic, only that microservice is scaled up.

This microservice concept underlies cloud-native. Combined with containers and orchestration, it allows teams to deliver new features faster and with less risk—**a perfect match** for DevOps principles and the near-constant innovation required by today's AI-driven solutions.

4.2 Containerization and Ephemeral Infrastructure

4.2.1 Containers vs. Virtual Machines

A core tenet of cloud-native is **containerization**:

- **Containers** (like Docker images) package up the application plus all its dependencies in a single, lightweight unit. This isolates the app from discrepancies in OS versions or library installations on the host.
- **Virtual machines (VMs)**, while providing isolation, are typically heavier—each VM includes an entire guest OS. Spinning up new VMs can be slower and more resource-intensive.

Containers are ephemeral: they can be started or stopped quickly, scaled horizontally under load, and replaced automatically if something fails. This ephemeral nature fits perfectly with continuous deployment—**each new build** can be spun up in test or staging environments, validated, then promoted to production without the overhead of traditional server provisioning.

4.2.2 Orchestration with Kubernetes

While containers are powerful, managing hundreds or thousands of them manually is impractical. Enter **Kubernetes** (K8s), the de facto standard for container orchestration:

- Automatically schedules containers onto available nodes (servers)
- Replaces or restarts containers if they crash
- Scales services up or down based on resource usage
- Manages networking and load balancing among containerized services

Kubernetes is typically run in the cloud (AWS EKS, Azure AKS, Google GKE) or on-prem via solutions like OpenShift. For DevOps teams, Kubernetes provides a **common, automated "layer"** so developers don't need to worry about the underlying machines. This abstraction fosters a **"platform"** mindset—teams deploy containers to K8s rather than dealing with server configurations. Combined with DevOps pipelines, code changes can trigger container builds, automatically tested, then orchestrated in production with minimal human intervention.

4.3 Infrastructure as Code (IaC)

4.3.1 Principles of IaC

Infrastructure as code moves away from manually provisioning and configuring servers, networks, and storage. Instead, every environment detail (e.g., how many instances, what type of load balancer, which security groups) is declared in **code**—like YAML for Kubernetes manifests or Terraform's HCL for cloud resources. The benefits include

- **Version Control**: Infrastructure definitions are stored in the same Git repos as application code, enabling reviews, rollbacks, and diffs.
- **Repeatability**: The same IaC template can create identical dev, test, or production environments.
- **Traceability**: Changes to infrastructure are tracked just like code commits, ensuring accountability.

4.3.2 Popular IaC Tools

- **Terraform by HashiCorp**: Cloud-agnostic, widely used to manage AWS, Azure, GCP, and other providers
- **AWS CloudFormation**: Native to AWS, using YAML/JSON templates
- **Azure Resource Manager (ARM)**: For defining Azure resources
- **Pulumi**: Uses general-purpose languages (TypeScript, Python, etc.) to define infrastructure

No matter the tool, the approach is consistent: write a declarative file describing "desired state" (e.g., "3 t3.medium instances, a load balancer, a VPC"), then apply it. The IaC engine ensures the actual infrastructure matches that state, updating or rolling back as needed.

4.3.3 Why IaC Complements DevOps

IaC merges well with CI/CD pipelines:

1. **Pull Request**: A developer changes a Terraform script to add a new microservice.

2. **Automated Plan**: The pipeline runs `Terraform plan`, generating a preview of changes (e.g., "create 2 new EC2 instances, update load balancer config").

3. **Review and Merge**: The team reviews and merges the PR if it looks correct.

4. **Apply**: The pipeline executes `Terraform apply`, provisioning the new infrastructure in a safe, trackable manner.

This synergy between DevOps and IaC helps organizations consistently spin up ephemeral test environments, replicate production conditions locally, and **tear down** resources once testing is complete—all automatically. The result is a faster, more controlled release cycle with minimal manual overhead.

4.4 Data-Centric Architectures and Observability

4.4.1 Breaking Down Siloed Data

We've seen how fragmentation can hamper end-to-end visibility. A **data-centric approach** ensures that from the moment code is committed to when it runs in production, **all relevant data** is captured in a unified manner—build logs, container metrics, test results, usage telemetry, security scans, etc. This means:

- **Consistent Tagging and Metadata**: For example, each container image is labeled with a commit hash, version number, and environment ID.

- **Centralized or Federated Logging**: Logs from containers, orchestration events, and infrastructure changes feed into a common system (Splunk, ELK stack, Datadog, etc.).

- **Unified Metrics**: CPU, memory usage, request latency, error rates—collected from all services for correlation.

- **Distributed Tracing**: For microservices, using tools like Jaeger or Zipkin to track requests across multiple hops.

Modern DevOps isn't just about code—it's about data. AI-powered observability tools ingest logs, traces, and metrics from thousands of sources, correlating anomalies faster than any human operator.

—Dynatrace, Moogsoft, and Splunk AIOps Solutions

4.4.2 Observability vs. Monitoring

Monitoring typically means collecting predefined metrics (CPU usage, memory, etc.) and setting alerts if they exceed thresholds. **Observability**, in contrast, is about **deep visibility** into system behavior. Observability solutions let you ask new questions on-the-fly (e.g., "Which microservice version correlates with a spike in errors?") without rewriting instrumentation. By designing your cloud-native stack with robust logging, metrics, and tracing from day one, you gain the ability to swiftly diagnose anomalies and feed that data into AI-driven analytics.

4.4.3 Real-Time Feedback Loops

A data-centric architecture also supports **real-time feedback** in DevOps:

- **Automated Canary Deployments**: Deploy a new service version to a subset of users; monitor error rates or latency. If metrics degrade, the system automatically rolls back.

- **Continuous Performance Testing**: Perform load tests on ephemeral environments, capturing metrics for regression analysis.

- **Anomaly detection**: Over time, ML-based solutions can watch normal patterns and flag suspicious deviations, accelerating root cause analysis.

Ultimately, the more comprehensive your data collection and correlation, the closer you get to **self-healing, self-optimizing** infrastructure—hallmarks of a NoOps future.

4.4.4 Platform-Agnostic Analytics

Even in a well-standardized, cloud-native environment, teams may use multiple solutions for source control, CI/CD, security scanning, test automation, and observability. A **leading** approach to unify analytics across these varied toolchains is offered by solutions like **Opsera**, which provides

- **Out-of-the-box integrations** with the majority of DevSecOps tools on the market

- A **platform-agnostic** way to aggregate build, test, and security data into unified dashboards

- **Standardized metrics** (like deployment frequency, lead time, and MTTR) regardless of the underlying CI or test framework

By consolidating data, leaders gain visibility into performance bottlenecks or security gaps across the pipeline—without forcing every team to adopt the exact same tool. This approach **reinforces** the cloud-native, data-centric architecture by **bridging** any remaining tool or data silos, enabling a truly holistic view of the software delivery lifecycle.

4.5 Putting It All Together: Integrated Cloud-Native Pipelines

4.5.1 A Typical Workflow

Imagine a developer merges a pull request to the `main` branch:

1. **CI Process**: A pipeline spins up ephemeral test environments using IaC (Terraform + Kubernetes).

2. **Automated Testing**: Unit, integration, and security scans run inside containers identical to production.

3. **Deployment**: If tests pass, the pipeline updates the Kubernetes deployment manifest, pinned to a new container image (`myapp:v1.3.5`).

4. **Observability Hooks**: Once deployed, logs, traces, and metrics feed into a central data store (e.g., Datadog, ELK)—potentially aggregated in a platform like Opsera for a unified, real-time view.

5. **Automated Rollout**: A canary release directs a portion of traffic to myapp:v1.3.5. If no anomalies are detected after a set window, all traffic shifts. If an error occurs, the pipeline automatically reverts to the previous stable version.

6. **Feedback**: Real-time dashboards, alerts, and notifications provide immediate data on success or failure. The developer sees consolidated logs and metrics in one place.

4.5.2 Security and Compliance in the Pipeline

DevSecOps means embedding security from the earliest stages:

- **SAST/DAST** (static/dynamic analysis) triggered on each commit or nightly.

- **Container security scans** checking base images for vulnerabilities.

- **Policy as code** to ensure resource configurations meet compliance (e.g., encryption at rest, restricted inbound ports).

- **Automated Gating**: If a high-severity vulnerability is found, the pipeline blocks deployment until patched.

By standardizing the cloud-native pipeline, you ensure these security checks are consistent and automated across all microservices, rather than applied sporadically or manually.

4.6 Case Study: Retail Giant Embracing Cloud-Native

A global retail company decided to overhaul its legacy monolithic ecommerce app. They spent years dealing with frequent downtime during holiday peaks and slow release cycles. Over 18 months, they

1. **Split the monolith** into microservices (checkout, product catalog, user profiles)

2. **Containerized** each service and adopted **Kubernetes** on AWS (EKS)

3. **Refactored** their manual server provisioning to **IaC** using Terraform, ensuring a uniform dev-staging-prod environment

4. **Implemented** a data lake approach for logs and metrics—every container logs to a central ELK cluster

5. **Matured** their pipeline with canary deployments and robust test automation

6. **Connected** it all with a standard DevOps platform, enforcing consistent tagging for each microservice and environment

Results:

- Deployment frequency jumped from monthly to **daily** in some areas.

- Outages due to scaling issues plummeted; Kubernetes handled surges.

CHAPTER 4 CLOUD-NATIVE AND DATA-CENTRIC APPROACHES

- Observability soared—teams could diagnose latency spikes in minutes by tracing a request through each microservice.

- Within a year, they began piloting **AI-driven** anomaly detection and auto-remediation, using the consistent data they now collected.

- Crucially, adopting a managed cloud service for their AI expansions meant **near-instant** access to new ML features and libraries, without lengthy on-prem upgrade cycles.

This journey exemplifies how adopting **cloud-native** technologies plus **data-centric** design transforms the software lifecycle. The firm's next step is exploring AI-based test generation and predictive capacity planning, building on their integrated pipeline. Notably, they stress how on-prem upgrades used to take **weeks** of planning—whereas new AI features in their cloud stack are now available **immediately**.

4.7 Why This Matters for AI and NoOps

4.7.1 Cloud-Native + Standardization = Data Gold Mine

Standardizing your cloud-native stack means every microservice runs in the same orchestrator, logs in the same format, and shares consistent metadata. This yields a **rich, uniform dataset** for AI algorithms to learn from. For instance, an AI system can see that "version v2.0.1 of service X tends to cause memory spikes after 5 hours, especially in region us-east-1." With enough data, the AI can **predict** or **prevent** incidents.

4.7.2 Autonomous Scaling and Self-Healing

Cloud-native architectures open the door for **autonomous scaling**—the system can automatically add pods or containers when loads rise, or kill them if underused, all without waiting for human intervention. By integrating AI, you can move from reactive auto-scaling to **proactive** or predictive scaling based on usage trends or event forecasts. Likewise, self-healing routines (e.g., auto-restart a failing pod, roll back a problematic deployment) become feasible at large scale because the orchestration platform can execute those instructions instantly.

4.7.3 Rapid Adoption of New AI Capabilities

One **often-overlooked** advantage of operating in a **cloud** (rather than on-prem) environment is the **instant availability of new features**—particularly relevant as AI evolves at record speed. AI platforms and ML services release new models, features, or frameworks frequently (e.g., updated large language models, advanced anomaly detection algorithms, or specialized GPU support).

- **On-prem upgrades** can be **expensive** and **time-consuming**, requiring new hardware, extended maintenance windows, and often weeks (or months) of planning.

- **Cloud-based solutions** can roll out new AI capabilities seamlessly. A platform update might instantly unlock the latest AI features—no major hardware refresh or lengthy downtime needed.

- This agility is a **quick win** for DevOps teams: as soon as an AI service or model is updated, the pipeline can integrate it in hours or days, keeping your organization on the cutting edge.

Combined with a **standardized toolchain**, these cloud-driven updates are easy to adopt across all microservices—accelerating your innovation cycle.

4.7.4 Developer in the Loop...For Now

We're not at complete NoOps yet—humans still design microservices, define IaC templates, and respond to novel incidents. However, as the pipeline collects more data and as AI agents mature, an increasing share of operational decisions can be **automated**. Ultimately, the **cloud-native, data-centric approach** is the runway on which AI "agents" can land, analyze, and take action. The ease of **continuously adding new AI features** in a cloud environment further propels this evolution toward minimal human oversight.

4.8 Key Takeaways and Next Steps

1. **Cloud-Native Essentials**
    - Embrace microservices for modular, independent deployments.
    - Containerize services for portability and rapid provisioning.
    - Use Kubernetes or another orchestrator for automated scaling, failover, and rolling updates.
2. **Infrastructure as Code**
    - Shift from manual server setups to declarative IaC for consistency, traceability, and speed.
    - Integrate IaC into your CI/CD pipeline for frictionless environment changes and ephemeral testing.

CHAPTER 4   CLOUD-NATIVE AND DATA-CENTRIC APPROACHES

3. **Data-Centric Observability**

    - Centralize logs, metrics, and traces with consistent tagging.

    - Enable real-time feedback loops (canaries, automated rollbacks) and advanced analytics.

    - Build a foundation for AI/ML to detect anomalies, predict resource needs, and eventually self-heal.

4. **Platform-Agnostic Analytics**

    - Solutions like **Opsera** provide out-of-the-box integration with most DevSecOps tools.

    - Deliver standardized dashboards for key DevOps KPIs (deployment frequency, lead time, MTTR), across diverse tech stacks.

    - Maintain tool flexibility without sacrificing centralized data insights.

5. **Instant Access to Evolving AI Features**

    - Cloud services offer **fast adoption** of new AI capabilities without the overhead of on-prem hardware upgrades.

    - This empowers DevOps teams to experiment and innovate rapidly, staying ahead of industry changes.

6. **Path Toward NoOps**

    - Standardized, cloud-native architectures feed the data needed by AI-driven automation.

    - Over time, more operational tasks—scaling, failover, config tuning—can be handed off to AI agents, freeing humans for higher-level innovation.

4.8.1 What's Next?

In the next chapter, we'll look at **what "good" truly looks like** by discussing reference architectures for DevOps: **one** integrated pipeline that unifies code, builds, tests, security scans, and environment provisioning. We'll also see how organizations can adopt **best-in-class** patterns without drowning in complexity—an essential step before diving into the generative AI transformations in subsequent chapters.

By blending **standardization** (Chapter 3) with **cloud-native, data-centric architectures** (this chapter), you position your DevOps environment for **scalability, reliability, and AI-driven innovation**. That's the recipe for the future of software delivery—and an essential milestone on the road to NoOps.

4.9 Chapter Summary

- **Definition of Cloud-Native**: Modern, microservices-based architectures with containers and orchestration, enabling fast, reliable deployments.

- **Infrastructure as Code**: Declarative, versioned environment definitions that integrate seamlessly with CI/CD for ephemeral and repeatable setups.

- **Data-Centric Approach**: Observability at every layer—logs, metrics, traces—to create real-time feedback loops and produce consistent data for AI.

- **Benefits**: Accelerated releases, improved reliability, dynamic scaling, **rapid adoption of evolving AI** features, and a standardized environment that fosters advanced automation.

- **Platform-Agnostic Analytics**: Tools like **Opsera** unify data from multiple DevSecOps solutions, delivering standardized metrics that enable consistent visibility and decision-making.

- **NoOps Outlook**: Cloud-native synergy with DevOps is the springboard for AI-based or autonomous operations. Over time, more tasks can be automated or predicted, reducing human toil and enabling teams to innovate faster.

CHAPTER 5

What "Good" Looks Like: A Reference Architecture

After exploring the **why** of standardization (Chapter 3) and the **how** of cloud-native, data-centric pipelines (Chapter 4), it's time to see what a **"good" architecture** actually looks like in practice. The goal here is to paint a clear picture of an **integrated, modern DevOps setup**—the kind that enables fast, reliable releases, robust security, continuous feedback, and a strong AI foundation. While no single blueprint applies to every organization, understanding the **reference patterns** that high-performing teams adopt can inform your own approach.

This chapter outlines a **model DevOps reference architecture** that unifies tooling, data flows, and organizational practices. We'll also highlight **key success factors** and "checkpoints" you can use to gauge whether your pipeline is truly delivering on the promise of DevOps—ultimately setting the stage for the generative AI transformations in subsequent chapters.

5.1 The Pillars of a "Good" DevOps Architecture

5.1.1 End-to-End Integration

A high-performing DevOps architecture places **every step** of the software lifecycle under one **coherent pipeline**: requirements, coding, building, testing, security scanning, deployment, and monitoring. Rather than multiple, disjointed pipelines or scripts scattered across teams, you have a **unified flow** that automatically hands off artifacts from one stage to the next. This ensures

- **Traceability**: From a single commit through build, test, and deploy, you always know *which* version of the code ended up where.

- **Consistency**: Each release candidate passes the same battery of checks (security scans, automated tests, compliance rules) before going to production.

- **Reduced Friction**: Developers and operators have **one** source of truth for where things live, how they're tested, and when they're deployed.

5.1.2 A Single Source of (Structured) Data

In Chapter 4, we stressed the importance of **data-centric** design. A "good" architecture ensures

- **All pipeline events** (commits, test results, vulnerabilities found, deployment statuses) flow into a **centralized data layer**.

- **Consistent Tagging**: Artifacts, logs, and metrics share metadata (e.g., commit hash, environment, microservice name) for easy correlation.

- **Real-Time Dashboards**: Leadership and teams can see accurate, up-to-date KPIs (deployment frequency, lead time, mean time to recovery, etc.) without manual curation.

5.1.3 Self-Service and Self-Healing

The best DevOps architectures are **self-service** for developers—meaning they can spin up new services, create pipelines, or run tests with minimal ops intervention. They also incorporate **self-healing** mechanisms, like

- **Automated Rollbacks**: If a canary deployment fails, revert to the previous stable version instantly.
- **Resilient Infrastructure**: Kubernetes or serverless platforms that handle failovers and restarts with no human input.
- **Automated responses** to certain incidents (e.g., scaling up memory if logs indicate an OOM risk).

5.1.4 Embedded Security and Compliance

DevSecOps is not an afterthought. A strong architecture bakes security and compliance checks into every phase:

- **Static analysis** on each commit; **dynamic testing** before production
- **Dependency scanning** to catch vulnerable libraries early
- **Policy as code** to enforce compliance gates for regulated environments

This approach turns security into a **continuous process**, rather than a dreaded final hurdle.

CHAPTER 5   WHAT "GOOD" LOOKS LIKE: A REFERENCE ARCHITECTURE

5.2 Reference Model Overview

Let's outline a common **reference model** that exemplifies these pillars in action:

1. **Requirements and Planning**

    - Teams track user stories, tasks, and features in a single system (e.g., JIRA, Azure Boards).

    - The same platform links to code repositories for traceability.

2. **Version Control and Code Collaboration**

    - A central Git-based repository (GitHub, GitLab, or Bitbucket).

    - Consistent branching strategy (e.g., trunk-based or GitFlow).

    - Pull requests/merge requests with automated code reviews and checks (linting, unit tests).

3. **Integrated IDEs**

    - A **standardized editor** or small set of approved IDEs, with **Visual Studio Code** as a prime example.

    - Preconfigured extensions or plug-ins (e.g., Docker, Kubernetes, linting, AI-based code suggestions like GitHub Copilot) for a **seamless developer flow**.

    - Built-in support for local builds/tests that match the pipeline environment, reducing "it works on my machine" issues.

CHAPTER 5   WHAT "GOOD" LOOKS LIKE: A REFERENCE ARCHITECTURE

4. **Automated Build and Test (CI)**

    - A dedicated CI engine (Jenkins, GitHub Actions, GitLab CI, etc.) that triggers on every commit or PR.

    - Containerized builds, ensuring reproducibility (Docker images for consistent environments).

    - Unit tests, integration tests, code coverage, static analysis, and security scans all run automatically.

5. **Artifact Management**

    - A repository for storing build artifacts or container images (e.g., JFrog Artifactory, Nexus, or Docker Registry).

    - Each artifact is versioned, tagged with metadata linking back to commits and issues.

6. **Infrastructure as Code (IaC)**

    - Automated provisioning of environments using Terraform, CloudFormation, or similar.

    - Staging and production environments that mimic each other's configurations.

    - Continuous integration for IaC changes (same PR ➤ plan ➤ apply flow).

7. **Integration and System Testing**

    - Beyond unit tests, advanced integration/system testing tools like **Functionaize** can validate end-to-end functionality.

    - These tests are triggered automatically in a staging or ephemeral environment, simulating real user flows.

8. **Continuous Delivery/Deployment (CD)**
    - Pipelines that deploy to dev/staging environments automatically on successful builds.
    - Canary or blue-green strategies for production.
    - Approval gates or automated tests that must pass before final deploy.

9. **Observability and Ops**
    - Centralized logging, metrics, and distributed tracing (Datadog, ELK, Splunk, etc.).
    - AI/ML-based anomaly detection, auto-scaling triggers, and auto-remediation playbooks.
    - ChatOps integrations (Slack, Teams) for real-time alerts and collaboration.

10. **Platform-Agnostic Analytics Layer**
    - Tools like **Opsera** that aggregate data from multiple DevSecOps solutions.
    - Standardized dashboards for deployment frequency, lead time, MTTR, etc.
    - Executive-level reporting plus granular views for dev/ops teams.

5.3 Example Workflow in Action

1. **Developer Creates a Feature Branch**
    - Picks a user story in the planning tool.
    - Branches off main using a naming convention (e.g., feature/USER-1234).

CHAPTER 5   WHAT "GOOD" LOOKS LIKE: A REFERENCE ARCHITECTURE

- Writes code in **Visual Studio Code** (or another standard IDE), benefiting from preconfigured linting, code formatting, and AI code suggestions.

2. **Pull Request**

   - On pushing to feature/USER-1234, the CI pipeline triggers automatically.

   - If the code passes linting, unit tests, static analysis, and any configured security checks, the PR is marked "green."

3. **Automated Integration Testing**

   - Merging the PR to main triggers a deeper test stage: container builds, integration tests, dynamic security scans, etc.

   - **Functionaize** (or other advanced test frameworks) runs system-level and end-to-end tests, validating real user journeys.

   - Artifacts get published to the registry if everything passes.

4. **Deployment to Staging**

   - The pipeline uses IaC definitions (Terraform) to spin up or update a staging environment.

   - Container orchestration (Kubernetes) deploys the new image with a canary rollout.

   - Monitoring data from logs and metrics is fed into an observability stack.

5. **Acceptance Tests and Security Validation**
   - Automated acceptance tests run in staging.
   - Additional security checks (e.g., penetration test scripts or advanced scanning).
   - If the pipeline sees anomalies or vulnerabilities, it **blocks** promotion.

6. **Promotion to Production**
   - On passing all gates, the pipeline triggers a **blue-green** or **canary** release in production.
   - The old version remains live until final validation. If metrics degrade, the pipeline instantly reverts.

7. **Postdeployment Analytics**
   - Tools like **Opsera** unify logs and metrics from the entire cycle, showing
     - Deployment frequency and success/failure rates
     - Test coverage and vulnerabilities detected/resolved
     - Real-time performance metrics for the newly deployed version
   - If issues arise, an AI-based anomaly detector may automatically open a ticket or roll back.

This entire workflow is designed to be **hands-off** unless a human is needed for approvals or to handle novel errors—an important stepping stone toward NoOps.

5.4 Organizational Design: The Supporting Structure

Even the best technical architecture falters if the **people and processes** aren't aligned. "Good" DevOps architectures typically go hand in hand with

1. **Cross-Functional Squads**

    - Each squad includes developers, QA, ops, and often a security champion.

    - They collectively own the pipeline, from code to production, preventing handoff silos.

2. **Platform/Center of Excellence Teams**

    - A specialized group that manages the shared DevOps platform (CI/CD tooling, IaC templates, monitoring stacks).

    - They build golden pipelines and reference architectures for squads to adopt.

3. **Mandatory Pipeline Requirements**

    - All code merges must pass automated tests, code scans, and compliance checks.

    - No shortcuts or hidden scripts outside the pipeline—this ensures a single source of truth.

4. **Continuous Learning Culture**

    - Postincident reviews with a blameless mindset.

    - Regular reviews of pipeline metrics and continuous improvement efforts.

- Shared knowledge about new AI capabilities or automation features.

With these organizational supports, you avoid pockets of fragmentation or "rogue pipelines" that undermine consistency.

5.5 Hallmarks of a Mature Reference Architecture

How do you know if your reference architecture is truly **"good"**? Here are some hallmark indicators:

1. **Minimal Manual Interventions**
   - The pipeline runs end to end without frequent manual steps.
   - Approvals are automated unless critical (compliance or major production changes).

2. **Rapid, Frequent Deployments**
   - Elite DevOps teams can deploy on demand or multiple times a day.
   - Lead time from commit to deploy is measured in hours (or minutes), not days or weeks.

3. **High Automated Test Coverage**
   - Unit, integration, security, and performance tests are mostly automated.
   - Builds that pass the pipeline rarely fail in production.

CHAPTER 5   WHAT "GOOD" LOOKS LIKE: A REFERENCE ARCHITECTURE

4. **Insightful Observability**
   - Teams see real-time dashboards showing health (CPU, latency, error rates), plus AI-driven predictions.
   - Root cause analyses can be done quickly, leveraging correlated data from logs/traces/metrics.

5. **Low Change Failure Rate**
   - A small percentage of changes require rollback or cause production incidents.
   - If failures do occur, the pipeline automatically reverts or fixes them with minimal downtime.

6. **Secure by Default**
   - Security scans and compliance checks are embedded, so vulnerabilities rarely slip through.
   - Encryption, policy compliance, and audit trails are standard pipeline features.

7. **AI-Enabled Automation**
   - The architecture's data richness fuels anomaly detection, predictive scaling, and even generative test creation (discussed in later chapters).
   - Over time, more ops tasks become autopilot, inching toward NoOps.

If you can check most of these boxes, you're well on your way to a best-in-class DevOps environment.

5.6 Real-World Example: A SaaS Company's Unified Pipeline

Imagine a mid-sized Software-as-a-Service (SaaS) provider with 20 microservices powering a collaboration platform. They adopt the reference architecture:

- **GitHub** for source control, using branch protection rules and mandatory PR reviews
- **GitHub Actions** for CI, with integrated security scans on every commit
- **Terraform and Kubernetes** in AWS for infrastructure and orchestration
- **Cloud-native** stacks for data (e.g., AWS RDS, DynamoDB) with automated backups
- **Visual Studio Code** as the main IDE, curated with official extensions for Docker, Kubernetes, and AI code suggestions
- **Functionaize** for advanced integration and system testing, automating end-to-end test flows for new features
- **Opsera** for unified dashboards and analytics across build pipelines, security scans, and production logs
- **Slack ChatOps** hooking into the pipeline for deploy notifications, automated incident creation

They structure squads by microservice, each owning its code, pipeline scripts, and IaC definitions. The **platform team** manages shared Terraform modules, container base images, and best-practice templates for new microservices.

Outcome:

- Deployments happen 10–15 times per day across the suite of microservices.
- On average, the lead time from code commit to production is under 2 hours.
- **Functionaize**-driven integration tests catch issues early, reducing production bugs.
- Security vulnerabilities are caught early—some squads measure an 80% drop in postrelease security tickets.
- AI-based anomaly detection identifies unusual spikes in usage, proactively scaling relevant microservices to maintain performance.

The company's leadership can track it all in a single analytics platform, ensuring no microservice lags behind. This reference architecture **empowers** each squad while preserving standardization and data consistency—exactly the balance DevOps aims for.

5.7 Common Pitfalls and How to Avoid Them

Even with a solid reference architecture, organizations can stumble:

1. **Partial Adoption**: Some teams bypass the pipeline for "urgent" fixes or shadow IT solutions. Over time, fragmentation reappears.
   - **Solution**: Enforce pipeline usage; disallow manual deployments or one-off scripts. Provide training for teams that struggle.

2. **Outdated Tests**: Automated tests degrade if not maintained. QA might rely too heavily on manual checks.

    - **Solution**: Make test coverage a KPI. Regularly refactor and update test suites. Tools like **Functionaize** help maintain robust integration suites.

3. **Too Many Exceptions**: A standard pipeline quickly becomes unwieldy if every team demands a custom workflow.

    - **Solution**: Allow small deviations only when justified. Maintain "golden pipeline" templates for 80–90% of use cases.

4. **Security/Compliance Gaps**: Relying on postdeployment audits instead of in-pipeline checks leaves vulnerabilities undiscovered.

    - **Solution**: Shift security left—embed scanning, policy checks, and compliance gates from the earliest build steps.

5. **Lack of Observability Investment**: Logging or metrics remain partial, siloed, or incomplete, hindering root cause analysis and AI adoption.

    - **Solution**: Treat observability as a first-class citizen. Budget time and resources to integrate logs, metrics, and tracing thoroughly.

Addressing these pitfalls ensures the reference architecture remains robust over time.

5.8 The Road Ahead

A **good** DevOps reference architecture doesn't stand still. As AI capabilities expand—particularly **generative AI** for coding, testing, and infrastructure—the pipeline will evolve to incorporate **automated code generation, self-healing scripts, and multiagent orchestration**. This architecture is the launchpad for deeper autonomy:

- **Generative AI for coding** (GitHub Copilot, etc.) plugging into the standardized IDE environment

- **AI-driven test orchestration** that automatically reorders or re-scopes tests based on code changes

- **Agentic AI** that can patch vulnerabilities, reconfigure infrastructure, or even spin up new microservices in response to user load, all while logging every action to your analytics layer

By adopting a **solid, standardized, cloud-native** reference architecture now, you prepare your teams for these emerging frontiers. The synergy of consistency, automation, and data-rich pipelines is precisely what generative AI needs to thrive in a DevOps context.

5.9 Chapter Summary

1. **Reference Architecture Pillars**
   - End-to-end integration, single source of structured data, self-service, embedded security.

2. **Concrete Workflow**
   - From planning to production, each step is automated and traceable, with minimal manual touches.

3. **Organizational Alignment**
   - Cross-functional squads plus a platform team ensure consistent adoption across microservices.

4. **Maturity Indicators**
   - Frequent deployments, minimal failures, automated tests, real-time observability, AI-powered automation.

5. **Common Pitfalls**
   - Partial adoption, old test suites, security/compliance bolted on too late, or ignoring observability.

6. **Future-Ready**
   - A robust reference architecture is the foundation for generative AI, autonomous agents, and eventually NoOps.

In the **next chapters**, we'll dive into **generative AI transformations**: how AI can supercharge coding, testing, infrastructure, and orchestration. But it all hinges on having a **reference architecture** like the one outlined here—a stable, standardized system that collects and correlates the data AI needs to make smart decisions.

Remember: the goal is **not** to force a single pipeline blueprint on everyone, but to provide **consistent guardrails**—the patterns and platform that make DevOps second nature. When these pieces come together, your teams move closer to **continuous innovation**, delivering user value at unprecedented speed and reliability. That's what "good" looks like.

5.10 Final Section (Part I): The Paved Road—Standardization, Cloud-Native Foundations, and Unified Insights

This closing section for Part I distills everything the reader has learned about killing toolchain chaos, embracing container-first infrastructure, and turning raw telemetry into board-ready KPIs. It introduces a fully opinionated "paved road" platform—GitHub + GHAS for code and security, Opsera for analytics, and VS Code as the single workspace—and then walks through a step-by-step implementation playbook (tiger team, pilot, rollout, decommission). By showing **how** to consolidate tooling, **how** to re-platform workloads, and **how** to wire velocity, risk, and cost into one dashboard, the section turns earlier concepts into a practical migration blueprint that readers can lift straight into their own organizations. In short, it is the bridge between strategy and execution—the launchpad for the NoOps future introduced in later parts of the book.

5.11 Executive Snapshot

Dashboards multiplied, logs scattered, security alerts hid in corners—and the business wondered why releases slowed. **Opsera** attacks the problem at the root: its integration fabric (83-plus native connectors covering SCM, CI/CD, cloud, IaC, observability, and testing) streams every build, scan, and deployment event into a single analytics layer called **Unified Insights**. That layer becomes the source of truth for DORA, SPACE, and cost dashboards, eliminating the swivel chair reporting that wastes engineer hours.

Around that hub, the paved road stack is deliberately lean. **GitHub Enterprise Cloud + Actions** handles code and pipelines, while **GitHub Advanced Security (GHAS)**—now unbundled so even Team plan orgs can buy *GitHub Secret Protection* and *GitHub Code Security* à la carte (effective April 1, 2025)—blocks leaked keys and vulnerable code inside every pull request. Container and Terraform templates ensure environments are *re-created*, never patched, so drift dies off naturally.

A single **VS Code** dev-container completes the experience: Copilot, GHAS SARIF viewer, and Opsera CLI are preinstalled, giving developers real-time feedback and KPI tagging the instant they open an editor. With lead time, MTTR, change failure rate, and tool count automatically harvested by Opsera at each migration milestone, platform engineering squads can prove value in weeks, decommission legacy licenses, and lay the clean telemetry foundation for autonomous NoOps.

Why executives care: The platform's simplicity slashes cognitive load; cloud SaaS delivery means upgrades—and new AI features—arrive automatically with zero downtime; and consolidating disparate scanners, CI servers, and monitoring tools into GitHub + Opsera cuts license spend while exposing real-time KPIs for every codebase, language, team, and business unit. Leadership can finally see lead time, MTTR, change failure rate, and cost trends in one place, act on bottlenecks immediately, and reinvest savings into innovation.

The result is measurable velocity, tighter security, reduced spend, and—most importantly—a clean telemetry foundation on which autonomous NoOps operations can thrive.

5.12 Key Takeaways

- **Place Opsera at the Analytical Core—Integrations First, Dashboards Second**: With 83-plus native connectors, Opsera harvests build, deploy, quality, security, and cost signals from the entire DevSecOps stack and then renders them in Unified Insights—the single pane of glass for DORA, SPACE, and 150+ KPIs.

- **Collapse Code, Pipeline, and Shift-Left Security into GitHub + GHAS**: GitHub Enterprise Cloud with Actions eliminates bespoke CI scripts, while the April 2025 unbundling of **GitHub Advanced Security** lets any org turn on **Secret Protection** and **Code Security**—blocking leaked keys and vulnerable code inside every pull request, even on the Team plan.

- **Adopt Container and IaC Templates So Environments Are Disposable**: Kubernetes service blueprints and Terraform modules recreate dev, staging, and prod on demand, eradicating drift and unlocking true cloud elasticity.

- **Treat Telemetry As a First-Class Artifact**: Enrich logs, metrics, and traces with `service/env/commit` tags via OpenTelemetry and stream them to Opsera; a single schema powers real-time dashboards today and AI analytics tomorrow.

- **Standardize on VS Code As the *One* Developer Workspace**: A vetted extension pack (GitHub Copilot, GHAS SARIF viewer, Opsera CLI) turns VS Code into the control point where coding, security scanning, telemetry tagging, and AI assistance all converge—ensuring every engineer starts in flow and every commit arrives fully contextualized for Unified Insights.

- **Run the Platform As a Product—Prove Value with Metrics**: A dedicated platform engineering squad pilots the golden pipeline, locks in adoption with branch protection rules, and uses Opsera to track lead time, MTTR, change failure rate, and license savings—turning standardization from aspiration into muscle memory.

5.13 Common Pitfalls

- **Best-of-Breed Tool Sprawl Rebuilds Silos—Opsera Becomes "Just Another Dashboard"**

    If teams keep their pet scanners, custom CI jobs, or niche observability SaaS, Opsera has to ingest from ten places instead of one, velocity data loses consistency, and license costs stay high. Standardization fails unless the *only* authoritative pipeline is GitHub Actions feeding Unified Insights.

- **Lift-and-Shift Without Re-platforming Traps You in Snowflake VMs**

    Fork-lifting legacy servers into a cloud VPC preserves brittle init scripts, kills auto-scaling, and often raises spend when 24 × 7 workloads meet on-demand pricing. Containers and IaC blueprints must replace pets with cattle before you can measure cloud efficiency in Opsera.

- **Telemetry Marooned in Point Solutions Starves AI and MTTR**

    Logs in one SaaS, metrics in another, traces nowhere: correlations break, MTTR stretches, and Copilot-for-Ops can't learn from fragmented history. Every signal must ride the OpenTelemetry ➤ Opsera path, enriched with `service/env/commit` tags, or Unified Insights becomes a partial view.

- **Security Bolted On at the Release Gate Breeds "Ticket Fatigue"**

    Running CodeQL, secret-scanning, and SBOM checks only in a late-stage environment floods backlogs and turns security into someone else's problem. GHAS (Secret Protection + Code Security) has to run in the pull request, and findings must flow straight into VS Code and Opsera, or developers will circumvent the process.

- **Ignoring the Single-IDE Mandate Fractures the Flow State**

    When a few squads stick with Eclipse, IntelliJ, or Vim, extension packs diverge, SARIF viewers disappear, and Copilot suggestions miss context. The VS Code workspace is the control point; bypass it and the golden telemetry tag set never appears in Opsera.

- **Partial Adoption Lets Drift Creep Back**

    One "urgent" hot-fix outside the golden GitHub workflow resurrects shell scripts, custom YAML, and rogue Helm charts. Without branch protection rules and Opsera compliance checks, standardization erodes in weeks.

5.14 Mitigation Playbook—From Strategy to Daily Habit

Consolidate the stack—curate, don't accrete. Define just four first-class pillars: **(1)** source and work management, **(2)** CI/CD, **(3)** observability/analytics, and **(4)** security.

- **GitHub Enterprise Cloud + Actions** = pillars 1 and 2.

- **Opsera Unified Insights** = pillar 3 (analytics surface + 80 + native connectors).

- **GHAS (Secret Protection + Code Security)** = pillar 4.

    One artifact repository (GitHub Packages or Artifactory) stores build outputs. Anything outside these pillars integrates *through* Opsera or OTLP—not as a parallel platform—cutting license spend and giving every commit, scan, and deployment a single addressable home.

Rationalize repos and pipelines.

- Rename repos to `<domain>-<service>`; switch every team to trunk-based branching.

- Publish a reusable `.github/workflows/release.yml` and reference it with `uses:` in each repo; updates propagate in one pull request.

- Turn on branch protection so nothing merges without the golden workflow and required GHAS checks.

Centralize telemetry before AI arrives.

- Deploy an **OpenTelemetry Collector** or language SDK beside each service; tag every signal with `repo`, `commit`, `service`, `env`.

CHAPTER 5   WHAT "GOOD" LOOKS LIKE: A REFERENCE ARCHITECTURE

- Stream logs, metrics, traces, test reports, cost data, and deployment events to **Opsera**. Unified Insights now lights up velocity, quality, cost, and *risk-per-release* dashboards—and begins stock-piling the clean training corpus future GenAI agents will need.

Shift security fully left—inside the pull request.

- Enable **Secret Protection (push protection)** and **Code Security (CodeQL, Dependabot)** org-wide.
- Findings surface in the repo's *Security* tab and the single IDE; fixes ship as PRs; pushes with secrets are rejected in real time.
- GHAS is now purchasable à la carte—even on GitHub Team—so cost objections disappear.

Wire speed, security, and spend into one timeline.

- Forward GHAS alerts, workflow statuses, and deployment outcomes to Opsera.
- Executives see live scorecards that correlate *lead time*, *MTTR*, *change failure rate*, and *license cost* with every release, service, and team.

Standardize the workspace, not just the pipeline.

- Mandate **VS Code** as the single IDE. Ship a vetted extension pack (Copilot, GHAS SARIF viewer, Opsera CLI).
- The template dev-environment guarantees identical compilers, linters, scanners, and telemetry hooks on every laptop or Codespace—so engineers stay "in flow" and every commit is automatically context-tagged for Unified Insights.

Guard against drift with lighthouse squads and policy as code.

1. Move a cross-functional pilot team onto the paved road end to end.
2. Capture before/after KPIs in Opsera; publicize wins.
3. Roll organization-wide: required status checks, scheduled compliance jobs, and Opsera dashboards that flag any repo lacking the golden workflow or IDE tag.

By eliminating tool sprawl, unifying telemetry, embedding security at the point of creation, and anchoring everything to a single VS Code workspace, you convert standardization from a slide-deck strategy into daily habit—and lay the self-healing, AI-ready foundation for NoOps.

| **Tool count reduction** | 22 → 7 | 7 | License inventory |

Hit these checkpoints and you migrate from legacy chaos to a unified, cloud-native, data-driven DevOps platform—complete with the clean telemetry foundation necessary for the NoOps era.

5.15 Implementation Guidance—Turning the Vision into an Org-Wide Upgrade Path

Below is a *repeatable playbook*: prove the paved road with a single squad, measure every move in **Opsera**, and then expand while you decommission the legacy jungle.

5.15.1 Quick-Start Checklist

1. **Stand-Up the Tiger Team:** Three to five platform engineers, one security lead, one SRE. Charter: design the paved road, define telemetry tags, run the pilot.

2. **Baseline the Sprawl:** Export every repo, pipeline, IDE plug-in, CI server, monitor, and scanner; snapshot DORA metrics and license spend in Opsera *before* changeover so impact is provable.

3. **Freeze New Tool Purchases:** All exceptions route through the tiger team.

4. **Pick the Core Stack**

    - GitHub Enterprise Cloud + Actions (source and CI/CD)
    - **GHAS** (Secret Protection + Code Security)
    - **Opsera Unified Insights** (analytics and connectors)
    - OpenTelemetry Collector (edge plumbing for high-volume logs/traces)
    - One artifact repo (GitHub Packages or Artifactory)

5. **Publish Repo and Tagging Standards:** <domain>-<service> naming, trunk-based branching, mandatory service/env/commit/ticket tags in every workflow.

6. **Roll Out the Single IDE:** Ship a curated **VS Code** workspace (dev-container or Codespace) preloaded with Copilot, GHAS SARIF viewer, and Opsera CLI. This is the control point where coding, scanning, and telemetry tagging begin.

7. **Launch the Pilot Codespace:** Verify secret-scanning and CodeQL warnings surface in-editor and are forwarded to Opsera.

5.15.2 Sequenced Migration Plan

Phase 0 (Weeks 0–2): Architecture and Proof of Concept

- Map current ➤ future state.
- Spin up a *sandbox* repo with the reusable release.yml workflow; stream logs/metrics/traces to Opsera via OpenTelemetry; push a dummy commit to prove end-to-end flow.

Phase 1 (Weeks 3–6): IDE and Git Standardization

- **Repo Rationalization:** Migrate fringe Git providers into GitHub; apply naming rules; enable branch protection.
- **Workspace Rollout:** Push the VS Code template to pilot squad; track *IDE Adoption* in Opsera (commits tagged ide=VS Code).
- **Retire Legacy Editors:** Remove corporate distribution of nonapproved IDEs and plug-ins.

Phase 2 (Weeks 7–10): CI/CD on GitHub Actions

- Convert Jenkins/Azure DevOps jobs into reusable workflows referenced with uses; register the wrapper job as a required status check.

- Tag every job with service/env; confirm build and deploy events populate Opsera velocity dashboards.
- Decommission Jenkins executors powering pilot services; reclaim VM budgets.

Phase 3 (Weeks 11–14): Shift-Left Security

- Enable Secret Protection (push protection) and CodeQL default setup org-wide; auto-enable Dependabot PRs.
- Surface findings in VS Code SARIF viewer; Opsera tracks *mean vulnerability remediation time* and *risk-per-release*.

Phase 4 (Weeks 15–18): Telemetry Consolidation and Legacy Sunset

- Deploy OpenTelemetry sidecars to the remaining services; backfill historic logs into Opsera.
- Decommission Splunk, New Relic, or custom log stacks once Opsera coverage ≥ 95%.
- Archive obsolete CI/CD repos; cancel surplus licenses.

Phase 5 (Weeks 19–24): Org-Wide Rollout and Guardrails

- Migrate two additional product lines per sprint—each inherits the VS Code workspace, golden workflow, and GHAS defaults.
- Enable scheduled compliance jobs; Opsera flags repos missing the golden workflow or IDE tag and auto-creates issues.
- Quarterly steering review: lead time, MTTR, change failure rate, tool count reduction, cost savings.

5.15.3 KPIs and Success Metrics (All Surfaced in Opsera)

Metric	Baseline	target	Measurement trigger
Lead time for change	–	↓ 30% by week 12	Commit → prod timestamp delta
Mean time to recovery (MTTR)	–	↓ 40% by week 18	Incident close in ops channel
Change failure rate	–	≤ 10% by week 18	Postdeploy hook result
Vulnerability remediation time	–	< 24 h median	GHAS alert → PR merge
IDE adoption (VS Code)	0%	100% pilot, ≥ 90% org	Commits tagged `ide=VS Code`
Golden workflow coverage	–	≥ 95% repos by week 18	Presence of `.github/workflows/release.yml`
GHAS coverage	–	100% repos by week 14	GHAS enabled flag
Container and IaC template adoption	–	≥ 90% services by week 20	Deploy events tagged `template=standard`
Product line migration progress	0 / N	+2 product lines per sprint	Migration tracker in Opsera
Telemetry coverage	50%	≥ 95% logs/metrics/traces	OTLP heartbeat per service
Tool count reduction	22	7 core tools	License inventory audit
Annual license savings	–	≥ $250 k	Finance feed → Opsera cost dashboard

CHAPTER 5 WHAT "GOOD" LOOKS LIKE: A REFERENCE ARCHITECTURE

These adoption metrics ensure you're not just improving speed and quality—you're proving that **the whole organization is actually using the paved road** and retiring the legacy jungle.

5.16 Glossary—Part I

- **DevOps**: A cultural and technical movement that unifies development and operations to deliver software faster, more reliably, and with continuous feedback loops.

- **Agile**: An iterative software delivery mindset (e.g., Scrum, XP) whose rapid sprints inspired DevOps to remove the "wall" between dev and ops.

- **Continuous Integration (CI)**: Practice of merging code to a shared branch multiple times per day, with automated builds and tests on every commit.

- **Continuous Delivery (CD)**: Automation that promotes every green build through test and staging all the way to production at the push of a button.

- **DevSecOps**: An evolution of DevOps that bakes security scanning and policy gates into every pipeline stage.

- **DataOps**: Applying DevOps principles to data-engineering pipelines so that datasets are versioned, tested, and delivered continuously.

- **MLOps**: Extends DevOps to the lifecycle of machine learning models (training, deployment, drift monitoring).

- **NoOps**: A vision of fully automated operations where infrastructure concerns fade behind self-managing platforms and services.

- **Standardization**: The disciplined reduction of tool sprawl and process variance to create a single "golden" pipeline and data schema.

- **Cloud-Native**: Architecting systems around microservices, containers, and managed cloud services so they can scale and heal automatically.

- **Microservice**: A small, independently deployable service that owns a narrowly scoped business capability.

- **Containerization (Docker)**: Packaging applications with all dependencies into lightweight images that run the same everywhere.

- **Kubernetes**: The de facto container orchestration platform that schedules, scales, and self-heals container workloads.

- **Infrastructure as Code (IaC)**: Declaring cloud resources (servers, networks, policies) in version-controlled files rather than clicking in a console.

- **Terraform**, **AWS CloudFormation**, **Pulumi**: Popular IaC tools that provision and update resources from declarative templates.

- **Ephemeral Infrastructure**: Short-lived, disposable environments that spin up on demand (e.g., for a test run) and are destroyed afterwards, preventing drift.

CHAPTER 5   WHAT "GOOD" LOOKS LIKE: A REFERENCE ARCHITECTURE

- **Observability:** A trio of logs, metrics, and distributed traces that provide deep insight into system health.

- **ELK Stack**, **Splunk**, **Datadog**, **OpenTelemetry**: Tooling that collects and visualizes those signals.

- **Git**: Distributed version control system underlying modern software collaboration.

- **GitHub**: Cloud platform for Git repositories plus pull request workflow, discussions, and marketplace integrations.

- Companion CI servers first referenced in Part I: **Jenkins**, **CircleCI**, **Bamboo**.

- **SonarQube/Snyk**: Static analysis and vulnerability scanning tools cited as common "shift-left" security steps.

- **Paved Road/Reference Architecture**: The opinionated, battle-tested pipeline, templates, and conventions every team can adopt out-of-the-box.

PART II

Generative AI Transformations

"Generative AI is no longer just a tool; it's a catalyst, reshaping code, tests, and the very craft of development in ways once unimaginable."

CHAPTER 6

Generative AI for Coding and Unit Testing

In the previous chapters, we established how a **standardized, cloud-native DevOps architecture** provides the ideal foundation for rapid, reliable software delivery. Now, we enter the domain of **generative AI**—a family of tools and approaches that can produce, refine, or transform content (including **code**). This chapter explores how **AI-driven coding assistants** and **AI-based test generation** can significantly boost developer productivity and code quality. While we'll dive deeper into AI for broader testing, security, and infrastructure in later chapters, the focus here is on **coding** and **unit testing**, where generative AI is already reshaping the developer experience.

6.1 The Rise of AI Coding Assistants

6.1.1 From Autocomplete to Intelligent Pair Programming

Developers are no strangers to **basic autocomplete**—IDEs have long suggested tokens, method names, or simple boilerplate. But in recent years, **large language models (LLMs)** trained on massive corpuses of public code have given rise to **AI coding assistants** that go far beyond old-school suggestions. Among these, **GitHub Copilot** stands out due to

- **Deep GitHub Integration**: Copilot is natively integrated with the GitHub ecosystem, automatically staying up-to-date with the latest code patterns and frameworks.

- **Seamless VS Code Experience**: It embeds directly into **Visual Studio Code**, offering in-context suggestions as developers type, so you always have the latest Copilot version without extra overhead.

- **Opportunity for Custom LLM Integration**: Organizations can extend Copilot or run custom models behind their firewall, tailoring suggestions to proprietary codebases or compliance requirements.

Other AI assistants (e.g., Amazon CodeWhisperer, Tabnine) also exist, but for our **reference architecture**, we'll focus on **GitHub Copilot** as a prime example of **advanced integration** with both GitHub and VS Code.

6.1.2 Why This Is a Game-Changer

AI coding assistants shift routine coding tasks—like writing boilerplate or searching Stack Overflow—off developers' plates. This can

- **Speed Up Development**: Freed from repetitive code, developers focus on design, logic, and complex problem-solving.

- **Improve Code Consistency**: The AI suggests patterns that adhere to recognized best practices (though these suggestions aren't always perfect, as we'll discuss).

- **Lower the Barrier to New Tech**: If you're unfamiliar with a particular library or framework, the AI can suggest usage examples on the fly.

- **Stay Always Updated**: Because Copilot leverages a continuously updated cloud service, you're always benefiting from the **latest** improvements to the underlying LLM.

Moreover, **platform synergy** plays a major role. While platform-independent AI tools can provide some value, **in-platform solutions** like GitHub Copilot often deliver a **deeper, more frictionless experience**:

- They have **first-class integration** with repository hosting, pull requests, issues, and CI/CD steps within the same ecosystem.

- The AI can more readily **tap into code context**, pull request history, or organizational coding patterns.

- As GitHub (or any integrated platform) evolves, the AI often gains **new features** and **seamless upgrades**—no separate licensing or installation overhead.

This **platform synergy** underscores how **strategic platform decisions** affect AI capabilities: adopting GitHub for code hosting and pipeline management, for example, can mean **instantly leveraging** Copilot across the entire DevOps toolchain.

When combined with a **standardized pipeline** (Chapter 5) and consistent data flows, Copilot neatly slots into daily workflows—especially in IDEs like **Visual Studio Code**—where the developer rarely leaves the editor to look up examples.

6.2 Generative AI in Practice: Coding Workflows

6.2.1 Prompting and Refining with GitHub Copilot

A typical workflow with GitHub Copilot might look like this:

1. **Comment or Prompt**: The developer writes a comment describing what they want ("Function that parses a CSV file and returns a list of dictionaries").

2. **Suggestion**: Copilot instantly generates a function snippet.

3. **Review and Edit**: The developer checks the logic, adjusts if needed, and merges it into the codebase.

Over time, Copilot learns from your coding patterns, the context in the file, and your acceptance/rejection of suggestions—thus **refining** its outputs. Some assistants also let you highlight existing code and ask for transformations (like converting from Python 2 to Python 3, or refactoring a big function into smaller ones).

6.2.2 Handling Edge Cases and Documentation

Copilot can produce **inline documentation** or docstrings explaining what the code does. It also attempts edge cases when generating code. However, developers remain responsible for verifying correctness—**AI can't guarantee** every corner case is handled. For mission-critical logic, you must still rely on thorough reviews and tests (unit, integration).

6.2.3 Team Collaboration and Code Reviews

Copilot suggestions typically appear at the individual developer's IDE level, but they can also be integrated into **pull request** workflows:

- **Automatic Code Review Comments**: GitHub (and other AI bots) can provide feedback in PR discussions, flagging potential bugs or style inconsistencies.
- **Suggested Refactors**: The AI might propose a simpler function signature or highlight repeated code blocks across modules.

This synergy helps busy teams maintain code quality even when reviewers are pressed for time.

6.3 Impact on Productivity and Code Quality

*Developers using AI coding assistants report a **10-30% productivity boost**, with improved accuracy and fewer defects, especially in repetitive coding tasks.*

—GitHub Copilot & AI Pair Programming (OpenAI & Microsoft)
Before vs. After—A Concrete Copilot Time-Saver

Example Scenario: A backend engineer must add OAuth 2 PKCE middleware to an existing TypeScript/Express API.

CHAPTER 6   GENERATIVE AI FOR CODING AND UNIT TESTING

Step	Manual workflow	GenAI-augmented workflow
1	Search documentation and Stack Overflow for PKCE examples	Type `// add OAuth2 PKCE middleware` comment
2	Write ~40 lines from scratch, tweak imports, handle errors	Copilot proposes 30 lines with error paths in <2 s
3	Manually add unit test scaffold	Precommit hook autogenerates Jest test skeleton
4	Run linter, fix six style warnings	Copilot code passes lint on first run
Elapsed time	≈42 minutes (incl. research)	**≈28 minutes** (−33%)

These numbers mirror GitHub's 2024 productivity study, where developers finished comparable tasks 30–47% faster with Copilot assistance.

Early adopters of GitHub Copilot and similar AI coding assistants report:

1. **Time Savings**

    - **Ten to thirty percent** faster coding for typical tasks. This figure can be higher for unfamiliar languages or frameworks, since the AI readily suggests patterns or libraries the developer might not know by heart.

2. **Reduced Cognitive Load**

    - Less time spent searching documentation or example code. The AI "front-loads" relevant snippets, letting developers remain in **flow** mode.

3. **Improved Consistency**

   - AI suggestions often follow standard patterns or style guidelines found in large code corpuses, reducing the chance of subtle bugs from copy-paste or ad hoc solutions.

4. **Rapid Onboarding**

   - New team members can rely on AI to fill in gaps or propose solutions that align with widely used practices, accelerating onboarding.

However, these benefits come with **caveats**—such as the risk of AI introducing insecure patterns or incorrect assumptions. Human oversight remains essential.

6.4 AI-Driven Unit Test Generation

6.4.1 Why Automated Test Creation?

Unit tests ensure that **low-level functions** work as intended. But writing them can be mundane—especially for boilerplate getters/setters, data transformations, or edge-case checks. GitHub Copilot, among other tools, can generate these tests automatically, freeing developers to focus on **more complex** or **domain-specific** testing scenarios.

> *AI-generated unit tests can achieve up to **70% test coverage** with **minimal manual intervention**, accelerating QA cycles and reducing developer effort.*
>
> —Diffblue Cover for Java Test Generation

6.4.2 Example Workflow with GitHub Copilot

1. **Code Changes**: A developer implements new functionality.

2. **Copilot Test Suggestions**: Using comments or prompts (e.g., "Write unit tests for function X"), Copilot suggests test cases right within VS Code.

3. **Review and Merge**: The developer inspects the generated tests, possibly adding or removing cases.

4. **Integration into CI**: Once approved, the new unit tests run automatically in the CI pipeline—just like any other test suite.

6.4.3 Benefits and Limitations

- **Benefits**: Quickly achieve higher coverage, reduce the burden of test scaffolding, and ensure a baseline of correctness.

- **Limitations**: AI-generated tests may not capture complex business logic or corner cases that require domain insight. Also, the tests rely on correct assumptions about how the code *should* behave. As always, humans must validate them.

6.5 Challenges and Limitations of Generative AI in Coding

Despite the significant upside, generative AI for coding has its **pitfalls**:

1. **Hallucinations or Incorrect Suggestions**

    - AI might produce code that **looks valid** but contains logic errors, insecure patterns, or references to nonexistent methods.
    - Always review suggested code before merging.

2. **Security Risks and Licensing**

    - Some AI suggestions might inadvertently reproduce copyrighted code from training data. Clarify your tool's licensing and policies.
    - AI code might also introduce vulnerabilities (e.g., SQL injection) if not carefully checked.

3. **Overreliance**

    - Juniors may rely heavily on AI-suggested code without truly understanding it, leading to knowledge gaps or decreased skill growth.
    - Encourage a healthy balance: use AI as an assistant, not a crutch.

4. **Context Limitations**

    - If the model lacks context about the entire system or up-to-date libraries, suggestions might be outdated or incomplete.
    - Copilot's continuous updates help, but thorough testing is still vital.

5. **Bias Toward Patterns**
    - AI is trained on popular code patterns found in public repos. This can perpetuate suboptimal designs if those patterns are widespread.

6.6 Best Practices for AI Coding and Unit Testing

1. **Human-in-the-Loop**
    - Don't blindly accept suggestions. Treat AI code as a **draft** that still needs your review and testing.
    - Encourage code reviews from peers or lead developers—even if AI "approved" it.

2. **Curate Prompts and Comments**
    - Write clear, descriptive comments or docstrings before requesting AI suggestions. The better the context, the more accurate the output.
    - For unit tests, specify the function's intended behavior or edge cases so the AI knows what to test.

3. **Integrate with CI**
    - Treat AI-generated code and tests like any other code: subject them to CI pipelines, linting, static analysis, and code coverage checks.
    - If a test is generated but fails consistently, investigate whether the code or test logic is at fault—don't just remove the test.

CHAPTER 6   GENERATIVE AI FOR CODING AND UNIT TESTING

4. **Security and Compliance**

    - Use SAST tools to scan AI-suggested code for vulnerabilities.

    - If your organization must comply with specific standards, ensure AI code meets those guidelines (e.g., cryptographic requirements, data handling).

5. **Education and Onboarding**

    - Provide training sessions for your team about AI coding best practices, pitfalls, and how to use GitHub Copilot effectively.

    - Pair junior developers with more experienced ones who can guide them on validating AI-suggested code.

6.7 The Road Toward Advanced AI-Driven Development

6.7.1 Evolution of Code Suggestions

Current AI coding assistants primarily rely on **text-based** deep learning models. However, the field is advancing toward

- **Context-Aware Models**: Systems that see your entire codebase or architecture, not just the current file, improving consistency

- **Multiagent Collaboration**: Different specialized AI agents that handle refactoring, performance optimization, or security analysis, working together with minimal human intervention

6.7.2 Unified Developer Experience

As these tools mature, we'll see deeper integration in **Visual Studio Code**, including

- **Real-Time Synergy with DevOps**: The AI can reference pipeline data (e.g., test coverage, recent bug reports) to shape suggestions.

- **Context from Production**: Observability data might inform the AI that a function is a hotspot for errors—leading it to propose more robust error handling.

6.7.3 Bridging to NoOps

When combined with robust DevOps, **AI coding and test generation** close the loop between development and operations. As code evolves quickly, unit tests follow suit automatically, ensuring reliability. Ultimately, fewer human interventions will be needed for routine tasks like debugging minor issues or writing boilerplate tests.

6.8 Chapter Summary

1. **GitHub Copilot As the Prime Example**

    - Offers real-time code suggestions in Visual Studio Code, seamlessly integrates with GitHub, and stays up-to-date via continuous cloud updates.

    - Supports both **coding** and **unit test generation**, enabling a more unified developer experience.

- Demonstrates **platform synergy**: in-platform AI often provides deeper, more frictionless integration than platform-agnostic tools.

2. **Generative AI Assistants**
   - Tools like Copilot transform coding from manual boilerplate to a guided, semiautomated process.
   - Productivity gains of **10–30%** are common, with higher developer focus on design and logic.

3. **AI-Driven Unit Testing**
   - Automated test creation raises coverage, catching simple regressions and freeing devs to focus on deeper logic.
   - Still requires human review for correctness and domain insights.

4. **Challenges and Best Practices**
   - AI can hallucinate or introduce flawed patterns; teams must maintain a human-in-the-loop approach.
   - Security and licensing considerations remain crucial.

5. **Future Directions**
   - Greater codebase awareness, multiagent systems, and full integration with pipeline metrics will push AI-driven coding closer to minimal human intervention.

- This sets the stage for advanced DevOps, where routine coding and testing tasks are **increasingly automated**—one more step toward NoOps.

In the **next chapters**, we'll examine how **generative AI** expands beyond coding—into **functional testing, integration testing, data and infrastructure** management, and eventually **pipeline orchestration**. By coupling AI-driven development with a standardized, data-rich DevOps environment, organizations can accelerate releases, raise quality, and reduce repetitive toil—laying yet another stepping stone toward a NoOps future.

CHAPTER 7

Generative AI for System and Integration Testing

In Chapter 6, we saw how generative AI can boost developer productivity by automatically suggesting code snippets and even generating unit tests. But testing doesn't stop at the function level. Modern software increasingly depends on **multiservice architectures**, dynamic user flows, and complex integrations—all of which require **functional and integration testing** to ensure a reliable end-to-end experience. This is where **AI-driven testing** tools—like **Functionaize**—come into play, offering advanced capabilities that automate or augment the creation, maintenance, and execution of more complex tests.

This chapter explores

- Why **functional and integration tests** are so critical in DevOps
- How **generative AI** can simplify or even **self-heal** test suites in response to changing applications
- The **practical workflows and best practices** for adopting AI-driven functional testing

CHAPTER 7   GENERATIVE AI FOR SYSTEM AND INTEGRATION TESTING

By the end, you'll see how AI transforms one of the biggest bottlenecks in software delivery—**comprehensive testing**—into a more seamless, automated process that supports continuous releases.

7.1 Why Functional and Integration Testing Matter

7.1.1 From Unit Tests to Real-World Scenarios

Unit tests confirm that **individual functions** or classes do what they should. But in a microservices or complex application world, these fine-grained checks aren't enough. Users rarely call a single method; they traverse entire workflows—logging in, browsing items, making purchases, etc. **Functional tests** replicate these real-world scenarios, ensuring the system behaves correctly from the **end-user's perspective**. Meanwhile, **integration tests** confirm that different modules, services, or APIs **interoperate** correctly.

7.1.2 The Pain of Manual Test Maintenance

Historically, writing and maintaining functional tests has been **labor-intensive**:

- **Scripting**: QA teams or developers spend hours scripting end-to-end flows in tools like Selenium, Cypress, or custom frameworks.

- **Frequent Breakages**: Minor UI or workflow changes break existing tests, leading to "flaky tests" that cause false positives or negatives.

- **High Maintenance Costs**: A large suite of test scripts can be time-consuming to update after each code or design change.

These challenges hamper agility—especially in a DevOps pipeline where new features drop frequently. Enter **generative AI**, which can **observe** applications, learn typical flows, and dynamically create or update tests as the application evolves.

7.2 The Rise of AI-Driven Functional Testing

7.2.1 Functionaize As a Prime Example

Numerous tools claim to apply AI to testing, but **Functionaize** stands out for its ability to

- **Automatically generate** functional and integration tests based on observed user flows or static analysis of the application

- **Self-heal** tests by dynamically adapting scripts when minor UI elements or pathways change

- Integrate deeply with DevOps pipelines—triggering tests on each build or environment change, capturing results, and feeding analytics back into a centralized dashboard

Other AI-based testing tools exist, but we'll use **Functionaize** as our primary example due to its notable emphasis on **advanced AI** features (like "model-based testing" and "anomaly detection" for test flows).

7.2.2 AI-Powered End-to-End Validation

AI-driven testing tools can do more than just **record and replay** user flows:

- **Heuristic or Model-Based Approaches**: The AI maps the application's possible states and transitions, discovering untested paths on its own.

- **Contextual Error Detection**: By analyzing typical user flows, the AI can spot anomalies or performance issues that might not appear in a straightforward script.

- **Automatic Test Updates**: When the application's DOM or API endpoints shift, AI recognizes the new structure and adapts the test steps, reducing flakiness.

The result is a functional or integration test suite that stays in sync with the evolving application, minimizing manual script rewrites.

7.3 AI-Enhanced Testing Workflows

7.3.1 Generating Tests

The workflow might look like this:

1. **Environment Setup**: In your dev or staging environment, Functionaize (or a similar AI test platform) observes typical user interactions or imports application specs (OpenAPI, user stories, etc.).

2. **Test Suggestion**: The AI suggests various end-to-end scenarios—login ➤ search ➤ add to cart ➤ checkout, for instance—based on recognized elements and flows.

3. **Developer/QA Validation:** QA or developers review these proposed tests, tweaking them for edge cases or specific validations.

4. **Automated Execution:** Tests run in CI whenever new code merges or a build is deployed to staging.

7.3.2 Self-Healing in Action

When the front-end changes (e.g., a button ID changes from btnCheckout to btnSubmitOrder), a traditional script might fail. However, an AI-driven approach can

- Identify that the button's position or text is similar to the old one

- Map it to the same semantic action ("checkout flow") and automatically update the test

- Log the change, allowing a QA engineer to confirm or override it

This self-healing minimizes test maintenance overhead—keeping your pipeline "green" more often and reducing manual rework.

7.3.3 Integration Testing Across Services

AI tools can also help with **API-level integration:**

- By analyzing **service definitions** (e.g., using Swagger/OpenAPI), the AI can propose end-to-end calls that chain multiple microservices.

- If a service interface changes slightly (new endpoint or parameter), AI-based integration tests can adapt.

- Coupled with environment data, the system might detect if certain microservices are down or misconfigured.

7.4 Benefits and Limitations of AI-Driven Functional Testing

7.4.1 Key Benefits

1. **Faster Test Creation**: Automating scenario generation accelerates coverage of real-world workflows.

2. **Reduced Maintenance**: Self-healing scripts adapt to small UI or API changes, cutting down on "churn" when features shift.

3. **Broader Coverage**: AI may discover flows that manual testers overlook, catching corner cases early.

4. **Better Feedback Loop**: Real-time updates in the pipeline mean devs see breakages quickly, aligning with DevOps principles of rapid iteration.

7.4.2 Challenges and Caveats

1. **Contextual Understanding**: AI can't always infer business rules or domain constraints. Manual validation of test logic is still necessary.

2. **False Positives/Negatives:** Self-healing might incorrectly map a changed element, or the tool might fail to detect a genuine bug if it interprets the new behavior as "expected."

3. **Complex Data Setups:** End-to-end tests often require seeded data, mock services, or orchestrated states. AI solutions can help with some of this, but advanced scenarios may still need manual setup.

4. **Performance Testing Gaps:** Most AI-based functional tools focus on correctness, not necessarily on performance or stress testing—those might require separate solutions.

7.5 Best Practices for Incorporating AI-Based Functional Testing

1. **Human-in-the-Loop Reviews**

    - Always review automatically generated test flows to ensure they align with **real business requirements**.

    - Approve or refine self-healing changes when the tool updates a locator or test step.

2. **Integration with CI/CD**

    - Run AI-driven functional tests in staging (or even ephemeral test environments) for every significant build.

    - Store results in a unified dashboard (e.g., Opsera or your chosen DevOps analytics layer).

3. **Use Version Control for Tests**

    - Even if the tool self-heals or auto-updates scripts, commit changes to a Git repo.

    - This ensures **auditability**—you can see exactly when and why a test changed.

4. **Combine with Observability**

    - If the AI flags a test flow as slow or flaky, cross-reference logs/metrics for anomalies.

    - Some advanced setups can feed production user paths to AI test tools, so the tests mirror real usage patterns.

5. **Training and Skill Building**

    - Teach QA engineers how to interpret AI suggestions, override incorrect assumptions, and shape complex test logic.

    - Encourage collaboration between devs and QA—functional tests are no longer just "QA's domain" if they're integrated in the pipeline.

7.6 Case Study: Ecommerce Platform Adopting Functionaize

A mid-sized ecommerce company with ten microservices (catalog, cart, checkout, user profiles, etc.) found their **end-to-end test suite** constantly breaking. Minor UI changes or new coupon flows caused failing scripts, leading to manual rework.

CHAPTER 7   GENERATIVE AI FOR SYSTEM AND INTEGRATION TESTING

1. **Implementing Functionaize**
   - They set up Functionaize in the staging environment, letting the AI **observe** user flows for a few sprints.
   - The tool generated **baseline tests** for login, browsing, adding to cart, and checkout flows—complete with validations for item details and pricing.

2. **Integration with CI**
   - On each build, the pipeline deployed to staging, then triggered Functionaize to run these tests.
   - If UI changes broke a locator, the tool auto-updated the script and flagged the modification for QA review.

3. **Outcomes**
   - The QA team reported a **60% reduction** in test maintenance overhead.
   - Critical paths (like checkout) had better coverage, catching edge cases (e.g., out-of-stock items or invalid coupon codes) earlier.
   - Developers felt more confident shipping updates daily, as the functional suite stayed green or provided quick, actionable failures.

4. **Future Plans**
   - The company plans to feed **production telemetry** into the AI, so it can adapt tests to real user behaviors (e.g., unusual multicurrency checkouts).

- They also want to incorporate performance checks, though that might require separate tools or advanced Functionaize add-ons.

7.7 The Road Ahead: AI Testing and the NoOps Vision

7.7.1 Beyond Scripts: Autonomous Test Agents

We're already seeing AI tools that attempt to discover **new flows** or **negative paths** without human guidance—akin to **agent-based exploration** of an application. Future solutions might

- Dynamically spin up test data or mocks
- Cross-verify logs, metrics, and application states in real time
- Collaborate with AI coding assistants to fix discovered bugs on the fly

7.7.2 Closing the Gap Between Dev, QA, and Ops

AI-driven testing aligns with the broader DevOps push toward **shared ownership**:

- Developers benefit from quick feedback when functional flows break.
- QA focuses on higher-level test strategy, letting the AI handle repetitive updates.
- Ops sees fewer false alarms from flaky tests, ensuring stable continuous deployment.

As AI matures, functional testing will become **less manual** and **more proactive**, bridging the gap between daily code commits and genuine user satisfaction.

7.7.3 OpenAI Operator: A Glimpse of Future System Testing

An illustrative example of **where AI-driven testing may evolve** is **OpenAI Operator**—an experimental AI agent that interacts with applications **exactly** as a human user would, via a built-in browser. Instead of calling APIs or using classic scripts, Operator navigates the UI, clicks buttons, fills forms, and reads on-screen text:

- **User-like Autonomy**: Operator interprets pages visually (thanks to GPT-4 with vision) and decides how to proceed based on test instructions given in **plain English** (e.g., "Apply discount code then complete the purchase"). It can handle multistep flows across different sites or services—much like a real user.

- **Business Logic Understanding**: Armed with GPT-4's extensive training, it can grasp high-level domain concepts ("expense reports," "best-selling products of 2022") and automatically figure out how to navigate the UI to achieve those goals.

- **No Traditional APIs**: Unlike standard integration tests, Operator doesn't call backend endpoints or rely on DOM locators. It's a **black box** approach that sees only what a user sees, verifying both UI and system behaviors together.

- **Reduced Maintenance**: Because Operator "visually" identifies interface elements, minor label or layout changes might not break tests. Like a human, it can adapt to small shifts or new page structures.

- **Challenges**: Currently, Operator can be immature, sometimes inconsistent, or blocked by certain protective features (CAPTCHAs, 2FA). It also can't handle sensitive tasks (like real banking transactions) without manual confirmation. Nevertheless, it showcases a potential future in which testers simply outline scenarios in natural language, and the AI autonomously handles the rest—**no code, no scripts, no direct API calls**.

If tools like Operator mature and integrate seamlessly into DevOps pipelines, they could push system and functional testing even closer to the **NoOps** vision—where AI agents handle routine test creation, execution, and adaptation, while humans focus on strategic test design and overall quality goals.

7.8 Chapter Summary

1. **Functional and Integration Tests**
    - Go beyond unit checks to validate end-to-end user flows and multiservice interactions.
    - Essential in microservice-based, user-centric apps.

2. **AI-Driven Testing (Functionaize Example)**
   - Automates scenario creation, adapts tests to UI or API changes, and integrates seamlessly with DevOps pipelines.
   - Significantly reduces the cost of test maintenance and ensures broader coverage.

3. **Key Benefits**
   - Self-healing scripts, rapid coverage of real-world flows, deeper insights, and continuous feedback loops.

4. **Limitations and Best Practices**
   - AI can't fully understand business logic without human guidance.
   - Manual reviews and acceptance remain critical.
   - Integrate with CI/CD, store tests in version control, and encourage QA-Dev collaboration.

5. **Future of AI Testing**
   - Emerging tools like **OpenAI Operator** demonstrate a UI-centric, user-like approach, potentially bypassing traditional APIs or element locators.
   - Could further reduce script maintenance, expand coverage, and bring system testing closer to a true "human-like" validation.
   - Another step toward NoOps, where routine QA tasks become largely automated, allowing teams to focus on innovation.

CHAPTER 7   GENERATIVE AI FOR SYSTEM AND INTEGRATION TESTING

Up next (Chapter 8), we'll see how **AI** also extends into **infrastructure provisioning**—writing or refining Terraform/CloudFormation scripts, optimizing configurations, and auto-healing IaC changes. This continuous automation from **coding ➤ testing ➤ infrastructure** elevates DevOps to a new level, drawing us ever nearer to the **NoOps** ideal.

CHAPTER 8

Generative AI for IaC and Data Provisioning

We've seen how **generative AI** elevates coding, unit testing, and functional testing. Now, we turn our attention to **infrastructure as code (IaC)**—an integral part of cloud-native DevOps (Chapter 4)—and extend that concept to **data provisioning** for realistic, secure testing environments. IaC ensures consistent, declarative, and version-controlled definitions of servers, containers, networks, and configurations. Meanwhile, data provisioning ensures that **test/staging environments** have accurate, representative data—often a subset or masked copy of production.

As environments grow more complex—spanning multiple cloud services, microservices, and compliance needs—managing both **infrastructure** and **test data** can become a bottleneck. In this chapter, we'll explore how **AI** can

1. **Generate** Terraform or CloudFormation scripts automatically

2. **Optimize** or **remediate** infrastructure configurations

3. **Provision** or **synthesize** masked production data for testing

4. Predict or **preemptively fix** misconfigurations and capacity issues

5. Enable a **self-healing** infrastructure that handles changes without constant manual oversight

6. Let teams **"stay in the flow"** by calling these actions via **natural language** (NLP) directly from the IDE

By blending AI with IaC **and** data management, organizations can further streamline their DevOps pipelines, cut down on errors, and push toward a **NoOps** reality where infrastructure and test data "just work."

8.1 Why AI for IaC and Data Provisioning?

8.1.1 Complexity and Rapid Changes

In modern DevOps, teams often juggle

- **Multiple environments** (dev, staging, prod) across different clouds
- **Microservices**, each with its own cluster of resources
- **Evolving compliance policies** (e.g., encryption, network segmentation)
- **Data privacy** considerations for test environments (masking PII, ensuring GDPR/CCPA compliance)
- **Frequent releases** requiring fresh datasets in test/staging to mirror production as closely as possible

Even with infrastructure-as-code tools (Terraform, CloudFormation, Pulumi), writing and maintaining large sets of configuration files can be tedious and error-prone. The same is true for **manually copying or anonymizing** production data to keep test environments current.

Generative AI can **assist** by analyzing existing infrastructure patterns, **suggesting** or **creating** new scripts, **automating** data provisioning tasks, and even **remediating** drift or noncompliant datasets.

8.1.2 Seamless Integration with DevOps and IDE NLP

Because IaC and data provisioning are version-controlled, any AI-generated changes can pass through the same **pull request** and **CI/CD** gates as application code. This synergy means

- **Developers or ops engineers** can review AI-generated Terraform or data-copying scripts in a PR, merging only after trust and compliance checks.

- The AI can integrate with existing DevOps dashboards, scanning real-time resource states, logs, metrics, and data compliance to propose changes.

- Teams can "**stay in the flow**" by issuing **NLP commands** directly from their IDE (e.g., Visual Studio Code): *"Create a masked copy of production data for the staging environment"* or *"Generate Terraform for a new microservice with an RDS MySQL instance."* The AI responds with proposed scripts or data workflows—**all without leaving the IDE**.

This loop—**monitor, propose, review, apply**—reduces the burden on operators and fosters a consistent, secure, and data-ready environment.

CHAPTER 8   GENERATIVE AI FOR IAC AND DATA PROVISIONING

8.2 AI-Driven IaC Generation and Data Provisioning

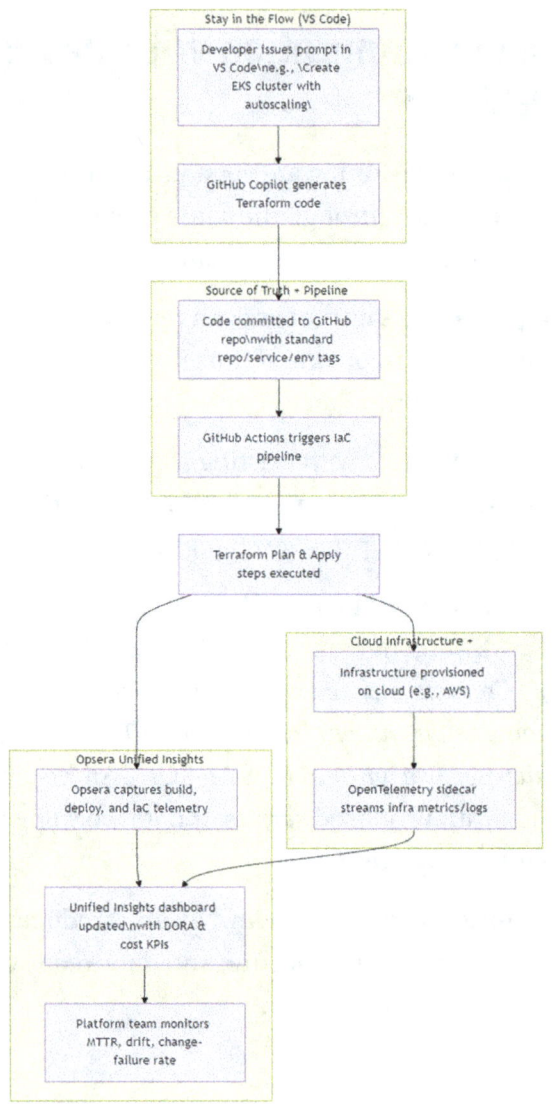

Figure 8-1. *AI-Driven IaC Workflow*

8.2.1 Automated Script Creation

Imagine starting a new microservice with certain infrastructure needs: an S3 bucket, a load balancer, auto-scaling groups, a database, and compliance tagging. Traditionally, you'd piece together Terraform modules from docs and examples. An AI solution can short-circuit this by

1. **Reading** a high-level description of your infrastructure requirements (e.g., "Highly available Node.js microservice in AWS with an ALB, RDS for data, auto-scaling EC2 instances, PCI compliance tags, and masked production data samples for QA testing")
2. **Generating** the corresponding Terraform or CloudFormation scripts—plus data-copying or anonymization routines if you want to replicate production data for test
3. Providing it in a **pull request** for human review

With minimal friction—**potentially triggered** via an NLP command inside the IDE—this approach speeds up creation of consistent, standardized infrastructure and data setups, especially for teams spinning up multiple services or requiring frequent test data refreshes.

> *AI can now **write Terraform scripts and YAML configurations automatically**, reducing infrastructure provisioning time by **over 60%**.*
>
> —Quali's Torque AI for Automated Provisioning

8.2.2 Refactoring and Modernization

Older IaC scripts can accumulate technical debt—hardcoded AMI IDs, outdated modules, or misconfigurations. Data pipelines for test environments might be manual or poorly documented. An AI can

- **Scan** existing scripts and data flows
- **Identify** outdated or redundant sections, or compliance gaps (e.g., unmasked user PII)
- **Propose** refactors (switching to new module versions, cleaning up unused resources) or an automated, masked data provisioning pipeline

This ensures your infrastructure and test data processes remain current with the latest best practices—similar to how AI code assistants auto-refactor application code.

8.3 Predictive Scaling, Drift Remediation, and Data Refresh

8.3.1 Predictive Scaling

One hallmark of **cloud-native** systems is the ability to scale dynamically based on load. But most scaling strategies rely on static thresholds or basic CPU/memory triggers. AI can take this further:

- **Analyzing** historical usage patterns (time-series data, user traffic)
- **Predicting** traffic surges (e.g., daily or weekly spikes)
- **Preemptively** scaling resources or adjusting auto-scaling rules for more efficient resource usage

Similarly, for data provisioning:

- The AI can predict **peak testing windows**, automatically refreshing or generating test data, ensuring that QA or staging always has the relevant dataset at the right time.

*AI-powered drift detection can **identify and self-correct** misconfigurations, ensuring compliance without manual intervention.*

—Firefly AI for Infrastructure Drift Detection

8.3.2 Drift and Misconfiguration Remediation

Infrastructure drift occurs when the **actual** environment deviates from IaC definitions—e.g., an engineer manually changes a security group in the AWS console. Data compliance drift might occur if new user data is left unmasked in a test environment. AI can

- Continuously **monitor** the live environment (infrastructure + data usage)
- **Compare** it with source definitions
- Either **notify** or **autocorrect** drift if rules allow, e.g., re-masking unencrypted data or reverting a misconfigured resource
- Propose or automatically enforce compliance for masked datasets

8.4 "Stay in the Flow": IDE-Centric, NLP-Driven Actions

A growing trend is enabling developers to **call all these actions** (infrastructure creation, data provisioning, environment refreshes) **directly from their IDE** using natural language:

1. **NLP Command Example**

   "Create a new staging environment in AWS with an RDS MySQL instance, a masked subset of production user data, and a load balancer for the Node.js microservice."

2. **AI Processing**

   - The AI understands the request, references your DevOps guidelines, compliance rules (e.g., PCI, HIPAA), and best-practice modules.

   - Generates Terraform or CloudFormation scripts, plus a data-masking pipeline.

3. **Pull Request**

   - The AI returns a proposed PR or changeset in your Git repo.

   - The developer or ops engineer reviews and merges once validated.

4. **CI/CD Application**

   - The pipeline spins up the environment, runs the data copy and anonymization job, and confirms readiness in a single go.

This approach reduces context switching—**developers remain in their IDE** to orchestrate not just code, but also infrastructure and data tasks, harnessing AI as an on-demand assistant.

8.5 Best Practices for AI-Driven IaC and Data Management

1. **Human Oversight**
   - Just like code suggestions, AI-suggested Terraform or data pipelines must be reviewed. Blindly applying changes can introduce hidden security issues or data exposure.
   - Maintain a **pull request** workflow where AI's changes are version-controlled and tested in staging before production.

2. **Security, Policy, and Data Compliance Checks**
   - Embed policy as code (e.g., Open Policy Agent) to ensure any AI-proposed changes meet compliance.
   - For test data, ensure the AI's instructions or generated scripts apply proper **masking** or anonymization steps to protect PII.
   - Keep a record of which agent or user triggered changes for audit.

3. **Clear Guardrails**
   - If your AI tool can auto-apply fixes or set up data, define which issues or tasks it can handle without human intervention.

- Start small—perhaps only auto-remediate drift or standard environment creation. Larger rearchitecture or major data migrations need manual approval.

4. **Continuous Learning**
   - Provide feedback to the AI if suggestions are incorrect or incomplete. Over time, it learns your naming conventions, resource usage patterns, data compliance rules, and so on.
   - Retain partial automation at first, gradually granting more autonomy as trust and accuracy improve.

5. **Cross-Functional Collaboration**
   - Dev, ops, and data governance teams should jointly define rules for AI-driven scaling, data provisioning frequency, and compliance parameters.
   - Keep the pipeline transparent so everyone knows *why* changes happen and *when* data is copied into a test environment.

8.6 Case Study: AI-Assisted Terraform and Data Masking at a FinTech Startup

A FinTech startup needed to rapidly spin up environments for new microservices, each requiring

- Secure VPC configurations
- RDS databases

- Strict PCI compliance (masking user payment info for nonprod)
- Regular data refresh from production to staging

1. **Initial Setup**
   - They introduced an AI-based IaC generator (built on a GPT-4 model) that took high-level specs (e.g., "Two-tier service with Node.js and RDS MySQL, requiring masked payment data for staging, PCI compliance tags").
   - The AI produced Terraform modules plus a data-masking pipeline script referencing the startup's existing "masking library."

2. **Review and Integration**
   - Ops engineers reviewed the AI's output in a Git pull request—verifying resource definitions, security group rules, compliance tagging, and correct data-masking parameters.
   - Merged the changes, triggering the CI pipeline to apply the Terraform in staging and run a job that copied + masked relevant user records from production.

3. **Remediation and Scaling**
   - Over time, the AI tool scanned for drift or resource inefficiencies (e.g., spotting underutilized staging RDS instances).
   - Proposed downsizing certain test environments off-hours.

- The tool also flagged data compliance drift when newly added user attributes weren't masked properly. Devs approved an auto-patch to rectify the pipeline.

4. **Outcomes**
    - New microservice environments were spun up **30–40% faster**, with consistent best practices for data security.
    - Drift and misconfigurations dropped significantly, as the AI regularly scanned and reported them.
    - The ops team had more bandwidth for strategic tasks—like advanced monitoring, performance tuning, and-deeper compliance audits.

8.7 The Road Ahead: Self-Healing Infrastructure and Data, Stay-in-Flow Approach

8.7.1 Multiagent Infrastructure and Data Management

In the future, we may see multiple specialized AI agents collaborating:

- One agent focusing on **cost optimization** (identifying underutilized resources or cheaper service tiers)
- Another on **security/compliance** (patching vulnerabilities, rotating credentials, ensuring masked datasets)
- A third on **performance/scaling** (proactive resource adjustments before load spikes)

- Yet another on **data integrity** (verifying anonymization rules, removing stale data, refreshing test sets on a schedule)

They coordinate changes through a single IaC + data pipeline source-of-truth, automatically creating PRs or applying fixes after threshold checks.

8.7.2 Operator-like Autonomy in Infrastructure and Data

Just as we see **OpenAI Operator** exploring apps from a user's perspective (Chapter 7), future infrastructure AIs might

- "See" a cluster's resource usage and the data usage patterns
- "Understand" new compliance rules or app usage
- "Act" within guardrails to keep resources healthy and data properly masked, possibly making corrections in real time without waiting on a human—**NoOps** style

8.7.3 NLP-Driven Flow in the IDE

The **"stay in the flow"** principle becomes more powerful as AI matures:

- Devs and ops open **Visual Studio Code**, or a similar IDE.
- They type or speak a **natural language command**: *"Spin up a new QA environment with partial masked data from production, and set auto-scaling to handle up to 1,000 concurrent users."*

- The AI interprets, references organizational policy, and returns a PR or direct pipeline action.

- The user confirms, or the pipeline auto-applies if the changes are within safe boundaries.

Gone are the days of switching between half a dozen consoles or writing hundreds of lines of Terraform and data scripts by hand. This integrated approach reduces friction and fosters near-instant environment creation.

8.7.4 Toward NoOps

If AI can autonomously

- **Spin up** new environments via simple NLP prompts
- **Optimize resource usage**, predict traffic, and scale accordingly
- **Patch misconfigurations** and drift
- **Enforce data compliance** by anonymizing or masking user data in test environments

the operational overhead shrinks drastically. Humans define goals, constraints, and policies but rarely intervene for day-to-day changes. This **NoOps** concept—where both **infrastructure** and **test data** are largely self-managing—becomes more tangible as AI capabilities expand.

8.8 Chapter Summary

1. **AI in IaC and Data Provisioning**
   - Generative AI can create and refactor Terraform/CloudFormation scripts, automatically provisioning or masking data for testing.

- Predictive scaling, drift remediation, and data compliance checks reduce manual overhead and error rates.

2. **Benefits**

    - **Faster provisioning** of new environments, with secure, representative datasets.

    - Automated or semiautomated **remediation** of misconfigurations, resource drift, and data compliance gaps.

    - **Predictive** resource scaling and scheduled data refresh, removing guesswork.

3. **Stay in the Flow**

    - NLP commands within the IDE let devs and ops trigger infra or data tasks without leaving their coding environment.

    - AI integrates with policy checks and PR workflows for safe, auditable changes.

4. **Challenges and Best Practices**

    - Human oversight remains crucial for security, compliance, and trust.

    - Clear guardrails ensure the AI doesn't apply harmful changes or leak sensitive data.

    - Feedback loops teach the AI your unique environment needs and data rules.

CHAPTER 8   GENERATIVE AI FOR IAC AND DATA PROVISIONING

5. **Case Study Lessons**
   - A FinTech startup used an AI-based IaC generator plus data-masking pipeline, reducing environment creation time by 30–40% and enforcing PCI compliance.

6. **NoOps Outlook**
   - Multiagent infrastructure + data management, real-time synergy with usage metrics, auto-scaling, auto-remediation, and NLP-driven creation.
   - Infrastructure and data provisioning become invisible overhead—**another major leap** toward NoOps.

In Chapter 9, we'll tackle **AI-orchestrated CI/CD**: how AI can optimize and adapt build pipelines, test sequences, and deployment strategies, from scheduling partial test suites to managing canary releases. Once combined with AI-driven coding, testing, infrastructure, and data provisioning, DevOps edges even closer to a **fully autonomous pipeline**, bridging us into the NoOps era.

CHAPTER 9

AI-Orchestrated CI/CD and Pipeline Optimization

We've seen how **generative AI** can revolutionize coding, testing, and infrastructure provisioning. Now, we turn to the **central nervous system** of DevOps: the continuous integration/continuous delivery (CI/CD) pipeline. A well-designed pipeline automates everything from **build** to **deployment**, but even robust pipelines can suffer from bottlenecks, flaky tests, slow feedback loops, or under-optimized release strategies.

This chapter explores how **AI** can

1. **Optimize** or **re-sequence** build and test stages
2. **Predict** failures and **suggest** or **auto-apply** corrective actions (e.g., partial test selection or canary rollouts)
3. Manage **deployment strategies** (blue-green, canary, rolling) dynamically

4. Integrate with real-time telemetry to **adapt** the pipeline on the fly

5. Let developers **"stay in the flow"**—issuing **NLP commands** directly from their IDE to orchestrate pipeline tasks

By injecting intelligence into the **CI/CD** process, teams can push code faster while maintaining reliability—another leap toward the **NoOps** future.

9.1 The Need for Smarter Pipelines

9.1.1 Complexity and Staging Bottlenecks

Modern pipelines often include

- Multiple test layers (unit, integration, performance, security)
- Various environment stages (dev, QA, staging, canary, production)
- Automated compliance gates (policy checks, risk assessments)

Each step can add **time** and **resource costs**. A monolithic pipeline that always runs *every* test or doesn't adapt to the code context can become a bottleneck. AI-driven orchestration can dynamically **rearrange** or **subset** test suites, tune concurrency, and schedule deployments more effectively.

9.1.2 Real-Time Feedback vs. Blind Scripts

Traditional CI/CD is mostly **script-based**—a linear or branched set of steps triggered on merges or scheduled events. If something goes wrong (e.g., a failing test), the pipeline halts; if everything passes, it proceeds. But it has **no real "intelligence"** to interpret logs, correlate issues, or propose deeper tests. AI can

- Observe test outcomes and logs in real time
- Identify patterns or anomalies
- Make decisions (e.g., skip certain tests if they're irrelevant to the changed code, or rerun tests it suspects are flaky)

This turns static scripts into **adaptive** pipelines.

CHAPTER 9   AI-ORCHESTRATED CI/CD AND PIPELINE OPTIMIZATION

9.2 AI-Driven Pipeline Optimization

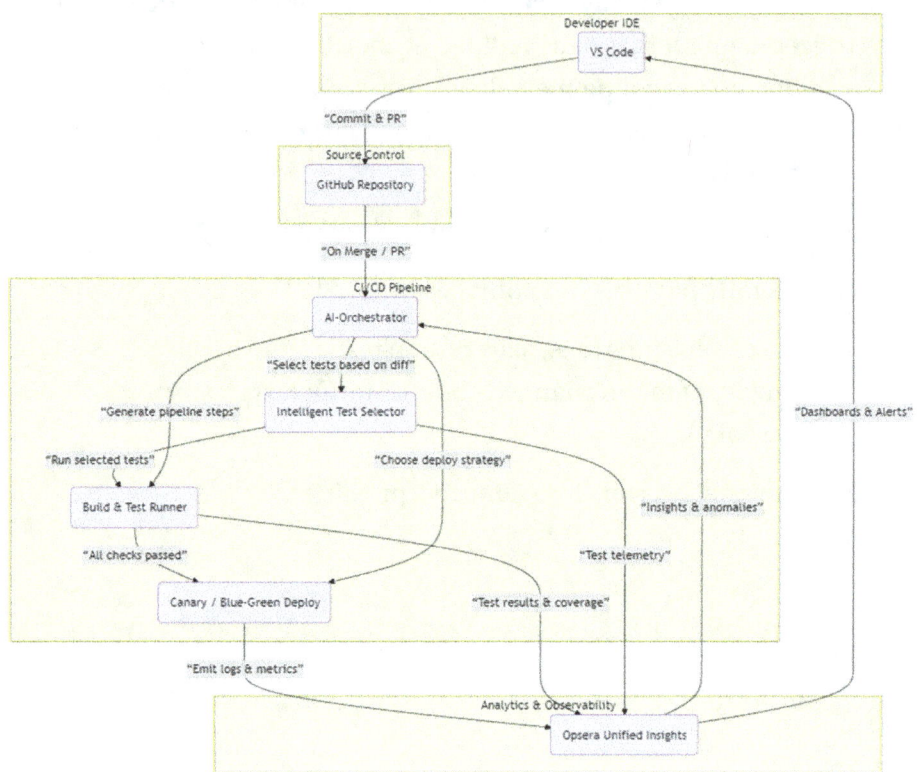

Figure 9-1. *AI-Orchestrated CI/CD*

9.2.1 Intelligent Test Selection

One of the biggest time sinks in CI/CD is running **all** tests regardless of what changed. AI can

1. **Analyze** code diffs, commit history, dependency graphs, or coverage data

2. **Choose** only the relevant subset of tests for that commit or PR (both functional and integration)

3. **Expand** the test set if it detects risky changes or anomalies

Result: Speedier feedback with minimal coverage sacrifice.

*AI-enhanced CI/CD pipelines can **predict failures before they occur**, dynamically adjust build steps, and optimize resource allocation for faster deployments.*

—AWS DevOps Guru & GCP Autopilot

9.2.2 Partial/On-Demand Deployment Sequences

Similarly, not every commit or change set warrants a full deployment to staging. AI can

- Trigger ephemeral environment creation or partial environment updates
- Decide to roll out a new feature only to canary or QA if it's a minor patch
- Defer certain resource-heavy steps (e.g., load tests) until the AI identifies higher risk

This fine-grained approach saves pipeline resources and ensures dev teams get feedback **sooner**.

9.3 Predictive Failure Analysis and Remediation

9.3.1 Anomaly Detection

As the pipeline runs, AI might

- Monitor logs, test results, code coverage, or infrastructure metrics in real time
- Detect suspicious patterns (e.g., a test suite that typically passes but suddenly fails on a small code change, or high CPU usage in a canary deploy)
- Flag potential root causes (maybe a newly introduced library version or a known vulnerability)

9.3.2 Auto-Apply Fixes or Reruns

When the AI identifies probable culprits—like a flaky test or a misconfigured environment—it can

- Retry the test with a known fix (e.g., increasing a timeout, cleaning up stale data)
- Automatically revert a problematic deploy if the error rate spikes
- Open a ticket or pull request with a recommended code or config fix, prompting a dev for final approval

In a more advanced scenario, the AI might even **apply** the fix if it's within safe guardrails, further reducing human toil.

9.4 Deploy and Release Strategy Optimization

9.4.1 Blue-Green, Canary, and Rolling

DevOps teams often pick one deployment strategy and stick to it. An AI-based pipeline can dynamically choose or adapt strategies per release:

- If changes are minor or low-risk, do a **rolling** deploy to production.
- If changes are high-risk or significantly alter performance, do a **canary** rollout first, sending a fraction of traffic to the new version and analyzing real-time user metrics.
- If quick rollback is essential, do a **blue-green** deployment for easy switching.

AI decides based on code diff risk, test outcomes, or historical data from previous changes.

9.4.2 Real-Time Telemetry Feedback

Once a new version is partially live, AI monitors logs (error rates, latency, user behaviors) to judge if the release is healthy:

- If metrics degrade, the pipeline auto-rolls back or reverts.
- If metrics improve or remain stable, the pipeline progressively increases traffic or finalizes the release.

Hence, the pipeline moves from a **scripted** approach to a **data-driven adaptive** approach, letting teams safely push changes more frequently.

9.5 Stay in the Flow: IDE-Centric, NLP-Driven CI/CD Control

9.5.1 Natural Language Triggers

With AI orchestration, developers can remain in their IDE and issue commands like

- *"Run the high-risk tests only on the checkout microservice for this commit."*
- *"Deploy this branch to canary with 10% traffic, watch for anomalies."*
- *"Rollback canary if error rate surpasses 2%."*

The AI interprets these requests, updates pipeline definitions or triggers ephemeral environment deploys, and surfaces real-time feedback—**all from the IDE**. No need to jump into a separate pipeline UI or manually edit YAML files.

9.5.2 Quick Feedback and Reduced Context Switching

Developers can see pipeline progress and logs inline—**Copilot-like** chat or status panels. If the AI detects anomalies, it can propose the following:

- *"We see the test coverage dropped 12%. Shall we run the entire suite or revert changes to maintain coverage thresholds?"*

Hence, **humans** stay in control but rely on AI for orchestration tasks, bridging code editing and pipeline management seamlessly.

9.6 Best Practices for AI-Orchestrated CI/CD

1. **Defining Risk Profiles**

    - Tag each microservice or code area with a **risk level**. AI uses this to decide how comprehensive tests or deployments should be.

    - Let the AI run fast, minimal tests for low-risk commits, and more exhaustive checks for high-risk areas.

2. **Guardrails and Policy**

    - As with AI for IaC, define what the AI can auto-apply. Full autonomy for canary rollbacks might be fine; major environment changes might still require sign-off.

    - Keep a robust audit trail—**which AI agent** made pipeline changes or triggered deployments?

3. **Train the AI**

    - Provide feedback or acceptance when suggestions or deployments are correct, and correct them when they're not. Over time, it learns your pipeline's specific patterns, test flakiness, and code risk profiles.

    - Periodically review how the AI classifies changes or schedules tests, refining or adjusting thresholds.

4. **Integrate Observability**

    - For advanced anomaly detection or rollback logic, feed real-time logs, metrics, and traces into the AI.

    - Summaries or anomalies appear in the IDE or pipeline dashboard, guiding devs and ops on next steps.

5. **Incremental Adoption**

    - Start with AI proposing partial test selections or canary rollouts, but require manual confirmation.

    - Gradually enable auto-apply for safe changes once you trust the AI's accuracy.

9.7 Case Study: Ecommerce Company's AI-Managed Pipeline

A mid-sized ecommerce platform with a large test suite struggled with 40-minute pipeline runs. They introduced an AI-orchestrated solution to cut times and reduce flakiness.

1. **Selective Test Execution**

    - The AI scanned each commit's code diff, identifying which microservices or modules changed.

    - It then triggered only the relevant test subsets (~30% of the entire suite on average).

    - Pipeline time dropped from ~40 minutes to ~20 minutes, and devs got feedback faster.

CHAPTER 9   AI-ORCHESTRATED CI/CD AND PIPELINE OPTIMIZATION

2. **Canary Deploy and Observability**

    - For major changes, the AI-orchestrated canary rollouts to ~10% traffic.

    - If error rates rose above a certain threshold, the pipeline auto-rolled back. No human action was needed except for final sign-off if a release was fully successful.

3. **NLP Commands in IDE**

    - Senior devs tested an integrated chat panel in VS Code, typing instructions like

    - *"Deploy feature/discount-codes to canary with 5% traffic for 2 hours."*

    - The pipeline recognized the request, updated the canary config, and then confirmed the schedule in a Slack message.

    - Postdeploy analytics were displayed inline, saving context switching overhead.

4. **Results**

    - Their pipeline was more adaptive, rarely running superfluous tests.

    - Production incidents due to bad releases declined ~40% as canary checks caught issues early.

    - Dev satisfaction improved—**82%** reported less frustration waiting for pipeline results, and more confidence in auto-rollbacks.

9.8 The Road Ahead: AI Pipeline Agents and NoOps

9.8.1 Multiagent Pipeline Collaboration

We can envision **dedicated AI agents** for

- **Test optimization** (selective runs, flake detection)
- **Deployment strategy** (blue-green vs. canary vs. rolling)
- **Security scanning** (inserting gates or auto-remediations if vulnerabilities are found)
- **Performance regression** checks (triggering load tests only when high-risk changes are detected)

They collaborate behind the scenes, each specialized in a dimension of the pipeline, with minimal human oversight—**NoOps** style.

9.8.2 Real-Time Observations and Automated Fixes

As we saw with Operator-like approaches (Chapter 7), future AI might not only detect anomalies but also **fix** pipeline scripts or environment variables on the fly—**like**

- "We see random test failures in environment X; shall we re-run them with an updated config or extended time-outs?"
- "Latency spiked after commit Y; rolling back canary from 30% to 10% traffic while we investigate."

This **autonomous orchestration** turns the pipeline into a living system that self-adjusts, merges or reverts changes, and continually optimizes resource usage.

9.8.3 NLP-Driven Flow from IDE

The **"stay in the flow"** approach extends fully to pipeline orchestration:

- Devs type: *"Create a partial test suite for payment microservice only and run it on staging. If pass, do canary at 5%."*
- The AI translates that into pipeline steps, enforces policies, and moves code through the pipeline.
- If everything is good, it pings the dev or auto-promotes to production. If not, it proposes rollbacks or additional tests.

This frictionless pipeline, guided by **human intent** but largely automated by AI, is the essence of **NoOps**: minimal manual steps, maximum automation, continuous intelligence.

9.9 Chapter Summary

1. **AI in CI/CD**
    - Generative AI can optimize build/test sequences, partial deployments, and dynamic rollouts.
    - Predictive failure analysis, auto-rollback, and anomaly detection reduce risk and accelerate feedback.

2. **Benefits**

    - **Faster pipelines** due to selective testing and dynamic concurrency.

    - Reduced production incidents via canary or rolling deploy with real-time AI monitoring.

    - **NLP integration** in the IDE fosters a "stay in the flow" approach—no separate UI or scripts needed for many pipeline tasks.

3. **Challenges and Best Practices**

    - Humans define guardrails—risk levels, test coverage minimums, canary thresholds.

    - Build trust incrementally, letting the AI propose changes but requiring manual acceptance at first.

    - Observability is key—AI needs metrics, logs, and test outcomes to decide.

4. **Case Study**

    - An ecommerce platform cut pipeline times ~50% by letting AI select relevant tests and orchestrate canary deploys. Incidents dropped, and developer satisfaction rose.

5. **Road to NoOps**

    - Multiagent pipeline collaboration, real-time environment adaptation, and NLP-driven orchestration from the IDE.

    - Pipelines become self-optimizing and self-healing, bridging the last gap to **fully autonomous** DevOps.

CHAPTER 9   AI-ORCHESTRATED CI/CD AND PIPELINE OPTIMIZATION

Up next—Chapter 10—we'll explore how **autonomous multiagent systems** unite all these AI capabilities (coding, testing, infrastructure, pipeline) into a cohesive, self-managing ecosystem, pushing DevOps ever closer to the **NoOps** dream.

9.10 Final Section (Part II): Catalyst to Autonomy—Generative AI Foundations for the Multiagent NoOps Era

This closing section for Part II distills everything the reader has learned about bringing large language model power directly into the developer workflow, turning brittle test suites into self-healing safety nets, and letting AI agents orchestrate infrastructure, data, and pipelines with near-zero friction. Like the "Paved Road" chapter 5.10 that capped Part I, it converts vision into a repeatable blueprint—only now the focus is moving from *standardization* to *autonomy*. GitHub Copilot (and future LLM plug-ins) sits inside the one-and-only VS Code workspace; Functionaize auto-generates and repairs functional tests; Opsera's Unified Insights captures every AI-driven build, scan, and deploy so leadership can watch DORA, SPACE, and *AI adoption* KPIs climb in real time. Follow the playbook here and your organization won't just *use* AI—it will embed it as muscle memory on the march to NoOps.

9.11 Executive Snapshot

Software delivery still stalls where humans grind through repetitive chores: writing boiler-plate code, hand-stitching test suites, tweaking Kubernetes manifests, chasing drift, and combing dashboards for anomalies. The

CHAPTER 9   AI-ORCHESTRATED CI/CD AND PIPELINE OPTIMIZATION

2025-era generative AI tools can now absorb *all* of that toil—*if* they are wired into a disciplined platform, measured against real KPIs, and governed by policy as code.

The AI-first paved road introduced in this section does exactly that.

1. **Code and Unit Tests Inside One IDE**
   GitHub Copilot operates in the standardized **VS Code** workspace, turning user stories into compile-ready code and companion unit tests while flagging insecure patterns before they land in the repo.

2. **System and Integration Testing That Heals Itself**
   Functionaize (or a comparable GenAI test engine) records true user flows, autogenerates functional tests, and self-updates when the UI or API shifts—eliminating the maintenance tax that cripples legacy test suites.

3. **AI-Generated Infrastructure That Never Drifts**
   IaC agents draft Terraform or CloudFormation modules on demand and feed them back into GitHub pull requests. A predictive scaling bot then fine-tunes cluster size ahead of traffic spikes, committing adjustments as code so nothing drifts in the dark.

4. **End-to-End Telemetry, Velocity, and ROI in One Lens**
   Every AI suggestion accepted, test healed, or drift patch applied is labelled with service/env/commit/source=AI and streamed to **Opsera Unified Insights**. Executives watch DORA, SPACE, *AI-generated LOC, self-healed test coverage*, and cost deltas rise or fall *in real time*—no spreadsheets, no swivel chair reconciliation.

5. **Security and Policy Guardrails by Default**
   GitHub Advanced Security (Secret Protection + Code Security) scans AI-generated code in the pull request, rejects leaked credentials in the IDE, and pipes findings straight into Opsera's risk dashboards. A policy-broker agent enforces what AI may auto-merge (typo fixes, infra drift under five lines) vs. what demands human eyes.

Why this matters right now.

- **Developer Throughput Soars**: Copilot accelerates feature delivery, and AI orchestration erases "pipeline busywork."

- **Quality and Resilience Climb**: Self-healing tests and drift bots close failure windows before users notice.

- **Security Shifts Even Further Left**: Issues blocked at the keyboard never reach production.

- **Costs Drop**: Obsolete CI runners, test frameworks, and monitoring silos get retired; cloud waste falls as predictive agents scale infra precisely.

- **Leadership Finally Sees AI ROI**: Unified Insights correlates every AI action with velocity, incident data, and dollars saved, turning hype into board-level evidence.

Standardization and cloud-native discipline from Parts I and II laid the runway; this section bolts on the AI engines that will lift the organization toward autonomous **NoOps** operations. Once these agents, guardrails, and metrics are in place, "keep the lights on" work becomes a side effect of the platform—freeing humans to focus on the next wave of innovation.

9.12 Key Takeaways

- **Treat AI Assistants As First-Class Teammates, Not Plug-Ins**: Wire Copilot, Functionaize, and IaC agents into the same VS Code dev-container that already enforces linting, secrets scanning, and telemetry tags.

- **Measure AI Adoption Early**: Track "AI-accepted code lines," "self-healed tests," and "auto-remediated infra drifts" alongside classic DORA metrics.

- **Keep the Analytics Core in Opsera**: Its 80-plus connectors turn AI events into board-ready KPIs without extra ETL.

- **Guardrails over Guesswork**: Define policy as code for what AI may *auto-merge* (typo fixes, drift patches) vs. what needs human review (schema changes, production data moves).

- **NLP Everywhere**: Let developers spin up masked data, run partial test suites, or trigger canary deploys by typing a sentence in the IDE—AI handles the YAML.

- **Run AI Enablement As a Platform Product**: A cross-functional "AI Guild" owns agent templates, prompt libraries, and success metrics, iterating just like any other internal platform.

9.13 Common Pitfalls

- **Shadow-AI Scripts Outside the Paved IDE**: If a team codes in IntelliJ with rogue Copilot settings, telemetry and security feedback disappear.

- **One-Off Prompt Engineering**: Ad hoc prompts create inconsistent code style and brittle test specs; without shared prompt libraries, AI output becomes the new tech-debt.

- **Untagged AI Activity**: Failing to log which code lines or infra commits were AI-generated blinds leadership to adoption rates and risk hotspots.

- **Overtrusting Hallucinations**: Accepting AI code without unit test coverage or SARIF scan resurrects the very defects automation promised to kill.

- **Drift Between AI Agents**: A pipeline bot might roll back a build the infra bot already scaled—unless a central policy broker arbitrates.

9.14 Mitigation Playbook—Hardening AI from Experiment to Everyday Muscle Memory

Goal: Convert generative AI potential into predictable productivity, security, and cost wins—without inviting drift, hallucinations, or shadow tooling.

CHAPTER 9   AI-ORCHESTRATED CI/CD AND PIPELINE OPTIMIZATION

9.14.1 Platform Guardrails

1. **Lock In the Single IDE (VS Code)**

    - **Action**: Publish an org-signed VS Code extension pack that auto-installs Copilot, GHAS SARIF Viewer, Opsera CLI, AI-prompt snippets, and your internal policy-as-code extension.

    - **Control**: Branch protection rule checks commits for ide=VS Code tag; nontagged commits fail CI.

    - **Win**: Guarantees every AI suggestion, scan result, and telemetry tag is generated, viewed, and logged in a uniform way.

2. **Centralize AI Telemetry**

    - **Action**: Extend your OpenTelemetry schema with ai_source, prompt_id, suggestion_accepted, self_healed=true/false, and confidence_score.

    - **Control**: GitHub Actions step rejects any merge lacking these tags.

    - **Win**: Unified Insights can correlate AI interventions with velocity, MTTR, and defect escape rate.

3. **Define a Policy-Broker Bot**

    - **Action**: Implement OPA/Rego or Cedar rules that classify AI changes:

    - **Green Lane** (auto-merge): Comment typo, doc update, infra drift patch < 5 LoC.

    - **Amber** (human review): Nonschema code, low-risk Terraform.

- **Red** (mandatory architect review): Data-schema change, security group rule, migration script.

- **Control**: Broker posts a PR label (ai-green, ai-amber, ai-red) and enforces matching review gates.

- **Win**: Keeps speed for the safe 80% while ring-fencing high-risk edits.

9.14.2 AI-Assisted Coding and Testing

1. **Copilot Coverage Mandate**

    - **Action**: Require every AI-generated function to come with Copilot-generated unit tests; GHAS blocks merge if coverage delta < +10%.

    - **Observation Hook**: Opsera board "AI Test Coverage Gain" = (AI functions w/ tests) ÷ (total AI functions).

2. **Shared Prompt Library and Style Guide**

    - **Action**: Store reusable, reviewed prompts for model fine-tuning (naming conventions, logging style, error patterns).

    - **Control**: A lint rule flags free-text prompts in code comments; suggests library equivalents.

    - **Win**: Consistent code style, fewer hallucinations, easier rollback.

3. **Self-Healing Test Pipeline**

    - **Action**: Integrate Functionaize (or similar) into nightly build; failed self-heals auto-open PRs with paired screenshots and diff commentary.

    - **Control**: QA triages via GitHub labels (auto-heal-accepted, auto-heal-declined).

    - **Win**: Functional coverage keeps pace with UI/API churn without manual upkeep.

9.14.3 Infrastructure and Operations

1. **AI IaC Generator with Two-Step Merge**

    - **Action**: IaC GPT writes Terraform in a feature branch; a static analysis action (tfsec, Checkov) + policy broker classify risk.

    - **Control**: Drifts < 5 LoC to existing module auto-merge (ai-green); larger changes require infra-review (ai-amber/ai-red).

    - **Win**: Mundane drift fixes commit themselves; architectural shifts still get eyeballs.

2. **Predictive Scaling Agent**

    - **Action**: Integrate KEDA/HPA rules suggested by an ML model; agent opens a PR every time forecasted traffic curve shifts threshold.

    - **Metric**: Opsera "Auto-scale savings" = (predicted capacity − actual utilization) × unit cost.

CHAPTER 9   AI-ORCHESTRATED CI/CD AND PIPELINE OPTIMIZATION

3. **Run-Book Copilot**

    - **Action**: Deploy a chat agent connected to RunDeck/PagerDuty APIs that can execute safe automations (ai-green) or draft playbooks for human approval (ai-amber).

    - **Win**: Compress MTTR without handing the keys to an unfettered bot.

9.14.4 Security and Compliance

1. **IDE-Level Secret Push Protection**

    - **Action**: Enforce GHAS push protection in VS Code; block secret commits before they reach the remote.

    - **Metric**: "Secrets Stopped at Keyboard" trend—should approach 100% vs. postcommit detections.

2. **AI-Aware SBOM and License Scan**

    - **Action**: Every accepted Copilot suggestion triggers a dependency sniff (SPDX, license text); GHAS fails PR on forbidden licenses.

    - **Win**: Closes the legal gap of unseen transitive dependencies.

3. **Continuous Policy Drift Audit**

    - **Action**: Nightly job compares live cloud config to IaC source; if drift > 5 LoC and **not** labelled ai-patch, raise critical alert.

    - **Win**: Prevents sneaky mis-config through side channels or mis-behaving agents.

9.14.5 Adoption and Business KPIs (All via Opsera)

KPI	Target	Alert threshold
AI-accepted LOC	≥ 25% total	‹ 10% 4-week slide
Self-healed test success rate	≥ 95%	‹ 80%
Mean vulnerability remediation time	‹ 24 h	› 48 h
MTTR (overall)	↓ 40% vs. baseline	Flat/rising
Drift auto-remediation coverage	≥ 90%	‹ 70%
Prompt library reuse	≥ 80% prompts	‹ 60%
Auto-scale savings	≥ 15% cloud spend	Flat/rising costs
License and tool count reduction	−15 tools, > $300 k	Savings plateau

Weekly Opsera dashboards color code each metric; quarterly steering reviews tie bonus pool funding to *AI value delivered*, not just "features shipped."

Summary: By combining *tight guardrails* (policy broker, GHAS, tagging) with *in-flow enablement* (VS Code pack, shared prompts, Functionaize) and *transparent ROI tracking* (Unified Insights KPIs), you ensure generative AI elevates velocity, quality, and security—instead of creating a new layer of chaos.

9.15 Implementation Guidance and Checklist—Turning AI Ambition into a Measurable Rollout

The playbook mirrors the structure used for the "Paved Road" Chapter 5.10 in Part I. One pilot squad proves the value, every action is logged to **Opsera Unified Insights**, and guardrails prevent drift as adoption scales.

9.15.1 Quick-Start Checklist

1. **Form the "AI Guild" Tiger Team**
   - Three senior engineers (dev, QA, SRE) + AppSec lead + FinOps analyst.
   - Mandate: own shared prompt library, policy broker, and AI KPIs.

2. **Baseline AI Readiness**
   - Capture today's DORA/SPACE metrics, test coverage %, mean vulnerability fix time, cloud utilization, and license spend in Opsera.
   - Inventory IDE diversity, test suite health, IaC maturity, and current Copilot usage (if any).

3. **Lock the Workspace**
   - Publish an org-signed **VS Code extension pack** (Copilot, GHAS SARIF viewer, Opsera CLI, policy-broker plug-in, shared prompt snippets).
   - Turn on a branch protection rule that rejects commits lacking the ide=VS Code tag.

4. **Wire Telemetry for AI**
    - Extend your OpenTelemetry spec with ai_source, prompt_id, suggestion_accepted, self_healed, and confidence_score.
    - Update the golden GitHub Actions workflow to fail if these tags are missing.

5. **Stand-Up the Policy Broker**
    - Deploy OPA/Rego (or Cedar) service that labels PRs ai-green, ai-amber, ai-red based on LoC, resource class, and security scope.
    - Configure required reviews that map to those labels.

6. **Connect AI Agents**
    - **GitHub Copilot**: Enable for pilot repo; enforce "unit test delta ≥ +10%".
    - **Functionaize**: Integrate via GitHub app; nightly self-heal job posts PRs.
    - **IaC GPT**: Enable via CLI wrapper that opens Terraform PRs with source=AI.
    - **Predictive Scaling Bot**: Tether to KEDA/HPA; writes PRs tagged ai-green.

7. **Sync with Opsera**
    - Verify AI tags, GHAS findings, test-heal events, Copilot acceptance logs, and infra drift PRs flow into Unified Insights dashboards.

9.15.2 Sequenced Migration Plan

Phase 0: Proof of Concept (Weeks 0–2)

Objectives

- AI Guild spins up a sandbox repository.
- GitHub Copilot, Functionaize, and the IaC-GPT wrapper each open dummy pull requests.
- Custom OpenTelemetry tags (`ai_source`, `prompt_id`, etc.) reach Opsera.

Key exit criterion

- A complete AI event is visible in Opsera Unified Insights (risk view).

Phase 1: Pilot Service (Weeks 3–6)

Objectives

- One product squad adopts the standard VS Code pack.
- Copilot is enabled on a 1 K-LOC microservice.
- Functionaize heals tests nightly; IaC-GPT patches drift; policy broker auto-merges *ai-green* PRs.

Key exit criteria

- AI-accepted LOC ≥ 15%.
- Self-healed tests cover at least 50% of that service's functional suite.

Phase 2: Code and Test Expansion (Weeks 7–10)

Objectives

- Enable Copilot organization-wide; enforce the *unit test delta* rule in CI.

- Publish prompt library v1 and add a linter that blocks undeclared ad hoc prompts.
- Extend Functionaize coverage to two additional services.

Key exit criteria

- Prompt library reuse reaches ≥ 60% of AI invocations.

Phase 3: Infrastructure and Operations (Weeks 11–14)
Objectives

- IaC-GPT activated for two platform teams.
- Predictive scaling bot manages nonproduction clusters; savings tracked in Opsera.
- Run-book Copilot integrates with PagerDuty (read-only mode).

Key exit criteria

- Drift auto-remediation ≥ 75% of detected drifts.
- Auto-scale savings ≥ 10% of cloud spend for pilot environments.

Phase 4: Org-Wide Security and Policy (Weeks 15–18)
Objectives

- Turn on GHAS push protection and CodeQL scanning in every repository.
- Enforce policy-broker label gating across the organization.
- Scheduled job flags commits without the ide=VS Code tag and opens remediation issues.

Key exit criteria

- GHAS coverage = 100% of repos.
- VS Code tag compliance ≥ 90% of commits.

Phase 5: Scale and Optimize (Weeks 19–24)
Objectives

- Migrate two additional product lines *per sprint* to the full AI stack.
- Hold a quarterly *AI Value Review* with CFO and CISO using Opsera dashboards.
- Retire overlapping test frameworks and legacy CI jobs.

Key exit criteria

- AI-accepted LOC ≥ 25% organization-wide.
- Annualized license savings ≥ $300 k.

9.15.3 KPIs and Success Metrics (All via Opsera Unified Insights)

Metric	Baseline	target	Trigger/query
AI-accepted LOC	0%	≥ 25% by week 24	Source=AI AND action=accepted
Self-healed functional tests	—	≥ 65% coverage	Functionaize healed=true events
Unit test coverage gain	—	+10% per PR	GH code coverage diff
Mean vulnerability remediation time	—	< 24 h	GHAS alert → PR merge

(*continued*)

Metric	Baseline	target	Trigger/query
Prompt library reuse	—	≥ 80% prompts	Opsera prompt-id histogram
Drift auto-remediation	—	≥ 90% infra drifts patched < 30 min	IaC GPT tag + diff size
Auto-scale savings	—	≥ 15% cloud spend	Cost Explorer feed vs. forecast
MTTR (all incidents)	—	↓ 40% vs. baseline	PagerDuty close time
License/tool count reduction	0	−15 tools, ≥ $300 k	Finance + inventory feed
IDE compliance (VS Code)	0%	≥ 90% commits	ide=VS Code tag presence
Policy-broker overrides	—	< 5% PRs	Broker label stats

Opsera dashboards surface each metric with trend arrows and SLA bands; weekly color-coded reports make slippage impossible to ignore.

Execution cadence

- **Daily**: Pilot squad stand-up reviews Copilot suggestion quality and test auto-heals.
- **Weekly**: AI Guild sync on prompt library, policy-broker exceptions, KPI deltas.
- **Monthly**: Org-wide demo day shows AI wins; finance and security update savings/risks.
- **Quarterly**: Steering committee ties bonus pool to AI Value Score (weighted average of KPIs above).

Follow this checklist and the organization will move from AI *experiments* to an **AI-amplified, self-healing NoOps reality**—with every gain captured in metrics the C-suite can trust.

9.16 Glossary—Part II

- **Generative AI/Large Language Model (LLM):** Models that produce code, tests, or prose from natural language prompts.

- **GitHub Copilot:** An LLM-powered coding assistant that surfaces whole functions, refactors, and unit tests inside Visual Studio Code.

- **Unit Test Generation:** Copilot (and similar tools) auto-write tests to raise coverage while developers focus on business logic.

- **Functionaize:** AI test automation platform that records real user journeys, generates functional and integration tests, and self-heals them when UIs change.

- **Self-Healing Tests:** AI updates locators or assertions when minor UI/API shifts would otherwise break scripted tests.

- **Visual Studio Code (VS Code):** The single, standard IDE in which Copilot suggestions, security scans, and pipeline commands converge.

- **Prompt Library:** A curated set of reusable prompts that keep AI output consistent with organizational style and policy.

- **GitHub Actions:** GitHub's native CI/CD runner that executes builds, tests, and deployments defined as YAML workflows.

- **GitHub Advanced Security (GHAS)**: Built-in secret protection, CodeQL scanning, and dependency-vulnerability checks integrated directly into pull requests.

- **CodeQL**: Static analysis engine underpinning GHAS that finds injection flaws, credential leaks, and insecure patterns in code.

- **Policy Broker:** An organizational gatekeeper that labels AI-generated pull requests (green/amber/red) and enforces who must review what.

- **IaC GPT**: A generative AI wrapper that drafts or refactors Terraform/CloudFormation modules from plain English intent.

- **Predictive Scaling Bot**: AI agent that studies traffic patterns and edits Kubernetes/KEDA/HPA settings ahead of load spikes.

- **Drift Remediation**: Automated detection and auto-patching of infrastructure that drifts from the IaC baseline.

- **Opsera Unified Insights**: A platform-agnostic analytics layer that aggregates build, test, security, and deploy events (including AI tags) into DORA/SPACE dashboards.

- **Canary/Blue-Green/Rolling Deployment**: Progressive release strategies that feed live metrics back to an AI orchestrator for go/rollback decisions.

- **Natural Language Pipeline Commands**: Typing "deploy to canary at 5%" in VS Code; the CI/CD agent translates and executes the request.

PART III

Multiagent AI and the NoOps Horizon

"True NoOps emerges when a constellation of specialized AI agents unites, orchestrating software delivery from end to end."

CHAPTER 10

Autonomous Multiagent Systems

> *The future of DevOps is not just automation, but **autonomous AI agents** that can plan, execute, and optimize entire software lifecycles.*
>
> —Futurum Research on Agentic AI (2025)

Across the previous chapters, we've explored how **generative AI** can enhance every stage of DevOps: coding, testing, infrastructure provisioning, and CI/CD orchestration. Now, we bring these strands together, envisioning a future where **autonomous agents**—each specialized in a particular facet of DevOps—work in concert to deliver **NoOps**: minimal human involvement in day-to-day operations. In this chapter, we'll outline

1. **What multiagent AI systems** look like in a DevOps context

2. **How these agents** collaborate to handle coding, testing, security, infrastructure, data, and pipeline orchestration

3. **Key benefits** (reduced toil, faster innovation) and **challenges** (trust, compliance, oversight)

4. A glimpse into the **NoOps** reality—a world where software changes practically manage themselves, letting humans focus on higher-value innovation.

10.1 Beyond Single AI Tools: The Multiagent Synergy

10.1.1 A Team of AI Specialists

Until now, we've mostly discussed AI in **one domain** at a time—e.g., GitHub Copilot for coding, Functionaize for testing, IaC generation tools, and pipeline orchestration AIs. In a multiagent world, each domain might have its own specialized AI agent:

- **Coding Agent**: Suggests/refactors code and unit tests, deeply integrated with the IDE

- **Functional Testing Agent**: Generates or self-heals integration and system tests (e.g., Functionaize)

- **Infrastructure Agent**: Proposes Terraform scripts, monitors drift, scales resources (Chapter 8)

- **CI/CD Orchestration Agent**: Adapts the pipeline, chooses deployment strategies, triggers partial test subsets (Chapter 9)

- **Security/Compliance Agent**: Continuously scans code, infra, data pipelines for vulnerabilities or policy violations, auto-remediating if allowed

- **Operator-like System Agent**: Possibly an AI that navigates applications as a user, verifying end-user flows (Chapter 7.7.3)

Rather than siloed tools, these agents can **interact** and **coordinate** through shared data or orchestrations. For instance, the CI/CD agent might consult the security agent before promoting a build, or the functional testing agent might request fresh masked data from the infrastructure agent. This synergy covers the entire DevOps lifecycle with minimal human intervention, as the AI "team" handles routine tasks.

10.1.2 Communication and Decision-Making

In such a multiagent system:

- Agents exchange **messages** or requests. For example, the coding agent might notify the CI/CD agent: *"I see major changes in the checkout microservice—please run advanced performance tests."*

- Agents share **context** (e.g., logs, test outcomes, environment metrics) through a central data layer or distributed event bus.

- Agents follow a **hierarchy or consensus** approach. Some organizations might designate a "master orchestrator" agent that finalizes decisions (like a pipeline orchestration AI), while others let each agent autonomously apply changes within its domain.

Ensuring they don't step on each other's toes requires **guardrails**—like role definitions, policy checks, and concurrency controls.

10.2 The Path to Autonomous NoOps

10.2.1 Fewer Manual Touchpoints

In a NoOps scenario, the day-to-day tasks we traditionally associate with operators—**provisioning servers**, **applying security patches**, **manually running test suites**—are handled by AI agents. Humans shift to

- **Defining high-level goals** (e.g., "We need 99.9% uptime," "We must mask PII in test environments," "We only allow certain container images")
- **Reviewing** or **approving** major changes or policy expansions
- **Investigating** novel incidents or edge cases beyond AI's knowledge

Everything else—routine releases, test updates, environment scaling—becomes an autonomous loop.

10.2.2 Intelligent Collaboration

Picture a new feature merged into `main`:

1. **Coding agent** (e.g., GitHub Copilot) might have helped write/refactor it.
2. **CI/CD agent** sees the merge, calculates risk, and triggers partial tests from the **functional testing agent** (like Functionaize).
3. If tests pass, the **infrastructure agent** spins up ephemeral environments if needed.
4. The **security agent** runs scans, ensures compliance, and signs off.

5. The **CI/CD agent** deploys canary or rolling updates, watching logs/metrics with help from an observability subagent.

6. If it detects high error rates or suspicious anomalies, it auto-rolls back or notifies dev.

7. If stable, traffic increments until fully live.

8. Meanwhile, data provisioning or ephemeral environment teardown is handled automatically once testing is complete.

No single step demanded a human push of a button or a manual script. Yet, the entire release cycle occurred seamlessly.

10.3 Benefits and Challenges of Multiagent Systems

10.3.1 Key Benefits

1. **Radical Efficiency**: Freed from daily ops tasks, teams focus on higher-level innovation, architecture, or user feedback.

2. **Consistency and Security**: Agents follow consistent rules and scripts, seldom forgetting best practices or skipping compliance checks.

3. **Scalability**: As organizations grow, adding more microservices or test suites doesn't drastically increase human toil—AI seamlessly handles more tasks in parallel.

4. **Resilience and Speed**: Agents can react in real time to issues—rolling back a failing deploy or patching a known vulnerability—often faster than a human on-call.

10.3.2 Challenges

1. **Trust and Oversight**: Granting AI autonomy means ensuring correct guardrails. A flawed or malicious suggestion from one agent could cause widespread issues if not caught by another.

2. **Policy and Ethical Boundaries**: At what point can an agent auto-commit code changes or manipulate production data? Organizations must define strict policies.

3. **AI Collaboration**: Agents must coordinate effectively. Otherwise, conflicting changes or concurrency issues can arise (e.g., one agent scaling up servers while another tries to tear them down).

4. **Training and Updating**: Each agent's ML model or knowledge base requires ongoing updates. They must remain current with new frameworks, cloud services, and organizational policies.

5. **Data Privacy and Security**: Multiple agents accessing code, logs, and data expansions raise questions about who can see what. Strict role definitions and encryption are essential.

10.4 Real-World Example: Toward an Integrated AI-Powered DevOps

Though fully multiagent NoOps is still emerging, some companies experiment with partial setups:

1. **AI Coding and Testing:** They use GitHub Copilot for code suggestions and an AI test generator for unit and integration tests.

2. **AI Infrastructure:** Terraform scripts are generated or refactored by an infrastructure AI, monitored for drift.

3. **AI CI/CD Orchestration:** Deployments run in partial auto mode with canary detection and rollback.

4. **Observability Hooks:** Real-time logs feed anomaly detection, which can trigger a pipeline revert or a new test run.

5. **Security/Compliance Agent:** Embedded scans block insecure merges or unmasked datasets in staging.

Over time, these pieces become more integrated, requiring fewer manual steps. Although humans still sign off on some changes, the system handles the bulk of routine DevOps tasks autonomously.

10.5 NLP and IDE Integration: "Stay in the Flow" for Everything

10.5.1 Unified Interface

In an advanced multiagent setup, a developer or ops engineer can launch or monitor these AI agents **directly from the IDE** (like VS Code), using natural language. For instance:

- *"Create a new QA environment for the payment microservice with masked production data. Deploy canary at 10% traffic if tests pass."*

- The multiagent system divides the request among the coding, infra, and CI/CD agents, orchestrating each step automatically.

- A final summary appears in the IDE's chat panel: *"QA environment created, data masked, canary deployed. Current error rate: 0.9%. Scaling traffic to 25%."*

10.5.2 Minimal Context Switching

This approach means devs and ops rarely jump to separate UIs or maintain scripts manually. They issue **high-level goals**, watch the pipeline's progress or logs in real time, and step in only if the AI requests confirmation. This fosters a **flow state**—less overhead, more creativity, and faster iteration.

10.6 Best Practices for Embracing Multiagent NoOps

1. **Incremental Adoption**

   - Start with one or two AI agents (coding/test generation, infra automation). Prove their reliability and build organizational trust.
   - Add more specialized agents over time, carefully defining roles and guardrails.

2. **Clear Guardrails and Policies**

   - Spell out which agent can auto-apply changes and which require sign-off. For instance, auto-remediation of small drifts is okay, but major refactoring or production data changes need a human check.
   - Employ policy-as-code (Open Policy Agent, etc.) and version control for all agent changes.

3. **Audit and Observability**

   - Log all agent actions with full context: who triggered it, which data or code was modified, and why.
   - Integrate anomaly detection not just in pipeline but across agent behaviors—detect any agent stuck in a loop or making repeated incorrect suggestions.

4. **Cross-Team Collaboration**

    - Dev, QA, security, and ops must define AI usage rules, risk levels, and compliance requirements.

    - Provide training on how to interpret or correct AI outputs, ensuring safe usage across the organization.

5. **Culture Shift**

    - Encourage teams to see AI agents as **collaborators**, not threats to their jobs. Emphasize how it reduces grunt work, letting humans focus on creative problem-solving and user value.

    - Communicate success stories widely to build confidence.

10.7 Looking Forward: The Emerging NoOps World

10.7.1 Ultimate State of Autonomy

In the full NoOps vision:

- Code changes flow from developer to production with near-zero manual steps—AI handles code suggestions, test creation, environment spin-ups, deployment decisions, and scaling.

- The system runs 24/7, auto-correcting issues and anomalies on the fly, only paging a human when novel or high-risk scenarios arise.

- Observability data feeds back into AI, constantly refining risk models, test coverage, and environment tuning.

10.7.2 Continued Role for Humans

NoOps doesn't mean **no** operators—it means operators' roles evolve:

- **Policy and Strategy**: Humans define the guidelines, compliance rules, and overarching goals.

- **Architecture and Innovation**: Humans dream up new features, design system topologies, and push the business forward.

- **Oversight and Ethics**: Humans ensure AI decisions align with ethical, legal, and organizational standards.

- **Incident Triage**: Humans handle the truly novel incidents that AI can't yet solve.

10.7.3 Constant Evolution

Multiagent AI systems require ongoing **learning** and **updates**. As frameworks, cloud services, and compliance standards evolve, the AI must keep pace—**just like** humans do. But each iteration brings us closer to an environment where software changes practically manage themselves, letting devs and ops focus on the future.

10.8 Chapter Summary

1. **Multiagent AI** in DevOps

   - Specialized agents for coding, testing, infra, CI/CD, security, and data provisioning, interacting to cover the entire lifecycle.

   - Communication and decision-making happen through shared data or orchestrations, with minimal human oversight.

2. **NoOps Vision**

   - Day-to-day tasks (provisioning, patching, test updates, scaling) become fully autonomous; humans only set high-level policies and handle exceptions.

   - Agents handle routine merges, canary rollouts, environment creation, and test results, often with real-time feedback loops.

3. **Benefits**

   - Massive efficiency, consistent best practices, rapid releases, fewer production incidents.

   - Humans focus on creative, strategic tasks—**the heart** of DevOps transformation.

4. **Challenges**

   - Building trust and oversight—guardrails, policy checks, audit logs.

CHAPTER 10 AUTONOMOUS MULTIAGENT SYSTEMS

- Training or updating multiple agents to keep them aligned and up to date with new technologies or compliance rules.

- Ensuring AI agents coordinate without conflict.

5. **Stay in the Flow**

    - NLP commands from the IDE unify dev and ops experiences, letting teams interact with AI agents directly, in context.

    - Minimizes context switching, fosters continuous collaboration, and catalyzes truly **seamless DevOps**.

6. **Road Forward**

    - Multiagent AI is the culmination of every automation and intelligence piece we've discussed: coding, testing, IaC, data provisioning, pipeline orchestration.

    - As these capabilities converge, DevOps enters a new era—one in which we see a **self-managing pipeline** that requires only high-level direction.

In Chapter 11, we'll step back and examine the **human–AI collaboration** factors—how roles change in a NoOps world, how to build trust in AI, and how to navigate the cultural shift required to embrace fully autonomous DevOps. Ultimately, NoOps is not about removing humans entirely but empowering them to **innovate** while the system handles the routine.

CHAPTER 11

Human–AI Collaboration

In previous chapters, we outlined how **AI** can transform DevOps practices—from coding, testing, infrastructure, and data provisioning to CI/CD orchestration—culminating in autonomous multiagent systems (Chapter 10). Yet the journey to a **NoOps** environment is not just about technology. It also demands a **cultural** and **organizational** shift in how humans work alongside AI agents.

This chapter focuses on

1. **The evolving roles** of developers, ops, QA, and security in an AI-powered, partially autonomous pipeline

2. **Building Trust and Managing Risk**: Best practices for human-in-the-loop oversight

3. **Upskilling and Team Dynamics**: How to help teams adapt to AI, from daily collaboration to new skill sets

4. **Ethical considerations** in letting AI make or influence critical decisions

Ultimately, NoOps is not about removing humans but **empowering** them to focus on higher-value innovation while AI handles routine tasks.

11.1 The Shifting Role of Humans in a NoOps Landscape

11.1.1 From Manual Operators to Automation Architects

Traditionally, operators or DevOps engineers spend large chunks of time on

- **Provisioning** servers, applying patches, performing routine deployments, etc.
- **Maintaining** or updating scripts for CI/CD, infrastructure, or data pipelines

In a NoOps scenario, these routine tasks are heavily automated by AI. Human operators become more like

- **Automation Architects**: Designing the guardrails, policies, and user stories that AI agents follow
- **Platform Curators**: Managing overarching platforms and AI frameworks, ensuring synergy across coding, testing, infra, and security
- **Strategic Problem-Solvers**: Investigating novel incidents or performance issues beyond AI's current understanding, orchestrating major redesigns or expansions

11.1.2 Developers As Product Creators

Developers, freed from writing boilerplate or dealing with pipeline friction, can

- **Focus on user needs**, domain logic, and architecture, letting AI handle trivial code suggestions or test generation
- Own **end-to-end** features, from code to production, but rarely push the buttons—AI coordinates merges, environment creation, and canary rollouts
- Interact with the pipeline or environment via **NLP** commands in their IDE—like *"Deploy this new feature to canary at 5% traffic"*

11.1.3 QA As Quality Engineers

In a NoOps setting, QA roles evolve:

- **AI Test Supervision**: Instead of writing endless scripts, QA engineers guide AI test agents (like Functionaize) to ensure coverage, define acceptance criteria, and refine self-healing test updates.
- **Quality Strategy and Exploratory Testing**: They focus on **strategic** test design, user journey mapping, and manual/creative exploration—areas where AI might lack domain context or empathy.
- **Data and Domain Expert**: QA ensures the AI's test approach truly reflects user expectations and business logic, clarifying edge cases the AI might overlook.

11.1.4 Security and Compliance Roles

Security professionals move from scanning or reacting to

- **Defining** policy as code (security baselines, compliance rules, data masking protocols) that AI enforces automatically.

- **Reviewing** AI's changes or recommendations to ensure no conflict with standards like PCI, HIPAA, and GDPR.

- **Incident Oversight**: If a security agent auto-patches or quarantines vulnerabilities, humans verify the appropriateness and handle severe incidents or advanced threat modeling.

11.2 Building Trust in AI
11.2.1 Human-in-the-Loop Approach

*While **81% of organizations** now use AI in DevOps, only **39% fully trust it**—proving that AI's success depends on transparency, oversight, and human-AI collaboration.*

—DORA's 2024 DevOps & AI Survey

A central principle is **human-in-the-loop** oversight, meaning AI suggestions or auto-actions typically flow through

1. **Proposal**: AI agent suggests a code fix, infra change, or test update.

2. **Review**: A dev/ops/QA/security person sees the proposal and checks it against the guidelines.

3. **Approval/Rejection**: If acceptable, it's merged; if not, the human corrects or modifies the AI's output.

4. **Feedback Loop**: The AI learns from acceptance or corrections, improving future suggestions.

Gradually, teams can grant the AI **wider** autonomy—auto-merging trivial changes or auto-rolling back canary fails—once they trust the AI's reliability in those domains.

11.2.2 Auditable Actions and Policy Checks

To maintain confidence:

- **Audit Logs Track Each AI Action**: Which agent made the change, the context, the outcome.

- **Policy Checks**: Tools like Open Policy Agent can instantly block any AI-driven config that violates security or naming conventions.

- **Regular Performance Evaluations**: Teams periodically review how many times the AI proposed changes, how many were correct, how many needed reverts or manual fixes—iterating on the AI's training or guardrails.

11.2.3 Transparency and Explainability

Developers, ops, and QA may be wary if the AI appears as a "black box," providing

- **Explanations** for suggestions or rollbacks ("I see a 15% error spike, so I'm reverting the deploy").

- **Context** about the data or patterns behind decisions, which helps users understand (and trust) AI actions. Over time, positive outcomes (e.g., AI preventing incidents) build a track record that fosters confidence.

11.3 Upskilling and Team Dynamics

11.3.1 Training Developers and Ops

In a NoOps future, humans must learn

- **AI Prompt Engineering**: How to effectively direct AI coding assistants, test tools, or pipeline commands in **NLP**

- **Policy and Governance**: Writing or maintaining policy as code that AI agents reference

- **AI Tooling**: Understanding the best ways to interpret AI logs, corrections, or proposals

These skills become as essential as Git or Docker knowledge once was. Corporate training or internal "AI champions" help spread these competencies.

11.3.2 Collaboration with AI Agents

Teams learn to treat AI as a **collaborator**:

- **Pair programming** with an AI coder (e.g., GitHub Copilot).
- **Co-review** pipelines or infra changes with an AI agent that proposes refactors.
- **Iterative** test design with a QA agent like Functionaize—humans specify scenarios and AI refines them.

This synergy reduces grunt work while keeping humans engaged in creative or strategic decisions.

11.3.3 New Roles and Leaner Teams

As the AI picks up routine tasks, the headcount needed for pure operational roles might drop—or those staff shift to more **value-driven** roles:

- Some organizations form a **Platform Engineering** or **Center of Excellence** team that curates AI usage, invests in training, and monitors agent performance.
- Others cross-train all devs and ops to become "AI-augmented DevOps engineers," each capable of controlling or guiding the AI for their domain.

11.4 Ethical and Compliance Considerations

11.4.1 Boundaries of AI Autonomy

When an AI can do everything from rolling back production to modifying infrastructure security rules, organizations must define

- **Critical actions** requiring human sign-off (like opening network ports to the Internet)
- **Sensitive data** or PII that AI should never access or replicate
- **Hard limits** on cost or resource expansions (no infinite scaling, for instance)

11.4.2 Bias and Reliability

AI might be trained on public code or standard best practices, which can embed biases or incomplete assumptions:

- A **coding agent** might lean on patterns from popular open source frameworks that are suboptimal for a specialized environment.
- A **testing agent** might prioritize mainstream user flows over niche use cases.

Teams must remain vigilant, ensuring domain-specific knowledge is integrated, and watch for any signs of harmful patterns or discriminatory outcomes (in data or logic).

11.4.3 Legal and Accountability

When AI makes or influences decisions that cause downtime or data breaches, who is responsible?

- Ultimately, the organization and the humans who configured the AI remain accountable—**NoOps** does not absolve accountability.

- Clear **governance** and **change management** processes ensure each agent's changes are traceable.

- Legal frameworks around AI usage in production are evolving, so compliance teams must stay updated.

11.5 Cultural and Organizational Shifts

11.5.1 Embracing AI As a Teammate

A big hurdle is **resistance** to new technology. Some fear job loss; others distrust AI suggestions. Leadership can

- **Communicate** that AI frees people from drudgery, letting them do higher-order tasks

- **Celebrate** successes (like an AI fix preventing an outage)

- **Reward** collaboration and AI usage, making it a positive, recognized activity

11.5.2 Learning from Failures

Even advanced AI can fail or produce flawed outputs. Encourage a **blameless** post-mortem culture:

- Analyze what went wrong, how the AI logic or training can improve, and how guardrails could prevent repeats.
- Avoid knee-jerk bans on AI after a single incident. Instead, refine its constraints, prompts, or policies.

11.5.3 Continuous Iteration on Roles and Processes

As AI capabilities expand, roles keep evolving. A developer might become more of a "product caretaker," or an ops engineer might pivot to "automation strategist." Regularly revisit **role definitions**, upskilling plans, and the division of labor between humans and AI. This fluid approach ensures the organization harnesses AI effectively rather than resisting it.

11.6 The Long-Term NoOps Vision

11.6.1 Humans As Strategic Overseers

In the end, NoOps envisions

- **AI agents** handling mundane tasks—monitoring health, scaling services, patching security holes, refreshing test data—at machine speed
- **Humans** focusing on creative pursuits: product roadmaps, user experience improvements, architecture decisions, and next-gen features

- A feedback loop in which humans define goals, AI implements them, and humans refine policies as new contexts emerge

11.6.2 Lifelong Learning and Evolving AI

NoOps is not static. Each day, the AI sees new commits, new incidents, and new performance data:

- It **learns** from each scenario, refining strategies for testing, deployments, or resource usage.
- Organizations also adapt, discovering new use cases or constraints for AI.
- The system becomes a living ecosystem of humans + AI co-creating software faster and more reliably than ever before.

11.6.3 The Human Touch

Even at peak autonomy, humans remain **essential**. AI might handle 95% of routine DevOps, but there will always be novel challenges—regulatory changes, business pivots, catastrophic incidents, or strategic leaps that require **human creativity**. The synergy is that humans and AI complement each other: the AI ensures operational excellence, while humans steer innovation, ethics, and big-picture direction.

11.7 Chapter Summary

1. **Evolving Roles**

   - Operators become **automation architects** and **platform curators**, while devs focus on product logic.

   - QA drives high-level test strategy, letting AI handle test generation and self-healing.

2. **Trust and Oversight**

   - A human-in-the-loop model ensures AI suggestions or auto-fixes pass through review.

   - Auditable logs, policy checks, and performance metrics help maintain confidence and accountability.

3. **Upskilling and Culture**

   - Teams must learn AI usage, prompt engineering, and policy definitions.

   - Emphasize AI as a collaborator, not a threat—success stories build acceptance.

4. **Ethical and Legal Factors**

   - Clearly define boundaries for AI autonomy and ensure sensitive actions remain guarded.

   - AI accountability falls under organizational governance—NoOps doesn't remove human responsibility.

5. **Organizational Shift**

    - Communicate the benefits (less toil, faster releases, higher quality) to overcome resistance.
    - Foster a blameless culture where AI mistakes lead to improvements, not bans.

6. **NoOps Future**

    - Humans as strategic overseers and AI as the operational backbone.
    - Continuous learning on both sides. The synergy pushes software evolution at unprecedented speed.

In Chapter 12, we'll explore **the future of software development**—extrapolating from AI-driven DevOps into a world where multiagent AI might **autonomously** generate entire features, compose test suites, and orchestrate everything. How will this shape the next 3–5 years, and what does it mean for the software industry as a whole? Let's find out.

CHAPTER 12

The Future of Software Development

Throughout this book, we've surveyed how **AI** is reshaping every layer of DevOps—from coding and testing to infrastructure provisioning and pipeline orchestration. We've seen how organizations can gradually adopt AI-assisted workflows, culminating in **multiagent systems** that push us toward a **NoOps** world, where day-to-day operations require minimal human intervention. Now, we look ahead to the **next 3–5 years** and beyond—extrapolating how emerging trends and technologies might further transform software delivery, developer roles, and organizational structures.

> By 2028, **75% of developers** will rely on AI-driven automation, fundamentally changing how software is built, tested, and deployed.
>
> —Gartner Hype Cycle for AI in Software Development

This chapter addresses

1. **Evolution of generative AI** in software development
2. **New frontiers** like full autonomous code generation, agentic collaboration, and dynamic user-driven development

3. **Implications** for the workforce, engineering education, and business strategy

4. **Reflections** on whether complete NoOps is truly attainable or always aspirational

By the end, you'll have a vision of where software is headed—and how to remain agile and competitive in a **rapidly shifting** tech landscape.

12.1 From DevOps to NoOps—What's Next?
12.1.1 Full Lifecycle AI

In the **NoOps** paradigm, we imagine AI

- **Generating** or suggesting high-level requirements
- Translating them into **architectural** designs and code scaffolding
- **Testing** new features and verifying security/compliance
- **Orchestrating** infrastructure spin-up, data provisioning, and continuous deployment
- **Monitoring** telemetry to spot anomalies or scale resources
- **Auto-resolving** routine issues and only escalating novel incidents

Current AI systems already tackle pieces of this. The **next wave** will see deeper integration and synergy across all domains, so that an organization's entire software lifecycle feels unified under an intelligent, adaptive umbrella.

12.1.2 Multiagent Collaboration at Scale

We increasingly see the promise of **agentic AI**—multiple specialized agents collaborating. For instance:

- A **requirements agent** that reads business tickets, converts them into user stories or acceptance criteria, and hands them off to...

- A **coding agent** that drafts code, test cases, or IaC definitions, then passes them to...

- A **testing agent** that refines functional or integration tests, verifying the new feature in ephemeral environments orchestrated by...

- An **infrastructure and pipeline agent** that configures the environment and automatically handles canary/blue-green rollouts.

Ultimately, humans focus on **high-level goals and creative solutions** while the AI "team" does the routine heavy lifting—**truly** bridging Dev, Ops, QA, and Security under one roof.

12.2 Autonomous Code Generation and Live Agentic Collaboration

12.2.1 Code As Conversation

As generative AI models grow more sophisticated, we may see entire features coded from **natural language** discussions:

1. **Product Manager**: *"We want a loyalty points feature for our e-commerce site, awarding X points per dollar, redeemable at checkout."*

2. **AI**: Interprets these high-level specs and scaffolds the code, tests, and data changes.

3. **Developers**: Validate the AI's design, refine logic or business rules, and reprompt if changes are needed.

4. **Pipeline**: Deploys the new feature on canary, checks user metrics, and rolls out fully if successful.

In short, "design by conversation," where code emerges as a byproduct of iterative, domain-focused dialogue between humans and AI.

12.2.2 Interactive Agents in the IDE

Tools like **GitHub Copilot** or other coding agents will likely evolve to entire "chatbot companions," not just autocompletes. These AI assistants

- Provide architecture diagrams, code, tests, and infra scripts in real time
- Use memory of your entire repo or even across services
- Collaborate with a functional testing agent or security agent behind the scenes for immediate feedback
- Potentially spawn ephemeral test environments or update user stories as the developer iterates

Essentially, your IDE becomes an **AI collaborator** station.

12.3 The Workforce and Organizational Impact

12.3.1 Upskilling and New Roles

As AI handles day-to-day tasks:

- **Engineers** pivot from manual scripting or debugging to **creative problem-solving**, policy creation, architectural thinking, user empathy, and bridging business needs with AI capabilities.

- **QA** focuses on advanced test strategies, domain logic, and user experience insights—backed by AI test generation for routine coverage.

- **Ops/DevOps** shifts to **platform engineering**, building internal frameworks for AI usage, policy enforcement, cost optimization, and security.

We also see roles like **"AI Ops Engineer"** or **"Data and AI Governance Lead"**—specialists who manage the interplay between humans and AI across the enterprise.

12.3.2 Leaner Teams, Faster Delivery

AI reduces some grunt work, meaning teams might be smaller or restructured. Skilled individuals can manage bigger, more complex systems because they delegate routine tasks to AI. This fosters faster delivery cycles and a capacity to **innovate** more often. However, it demands **cultural acceptance**—some organizations might resist changing roles or trusting AI for critical tasks.

12.4 Business Strategy and Competitive Advantage

12.4.1 Time to Market and Continuous Innovation

Companies that master AI-driven DevOps can release features rapidly and reliably, seizing market opportunities before competitors. They can

- Iterate on product ideas in days instead of months
- Continuously test new user flows or microservices with minimal overhead
- Reap cost savings by auto-optimizing infrastructure usage

This "**ultra-agile**" capability can differentiate winners in a saturated market.

12.4.2 Data Monetization and AI Feedback Loops

In a NoOps environment, the system constantly collects telemetry, user behaviors, and logs. AI agents can

- **Derive** insights for new features or performance enhancements
- **Close the loop** by adapting systems automatically, or exporting data to business intelligence
- Potentially drive new business models or personalization strategies (e.g., dynamic user experiences, targeted features)

Organizations that harness AI's feedback loops more effectively might shape new product lines or revenue streams.

12.5 Challenges and Limitations in the Emerging NoOps Era

12.5.1 Complexity and Interagent Conflicts

As multiagent solutions grow, complexity can become daunting:

- Agents might conflict—one scaling up resources, another trying to reduce costs.
- Coordination frameworks are required, or a "meta-agent" orchestrates final decisions.
- Debugging agent behaviors can be nontrivial if you don't have transparent logs or a robust policy layer.

12.5.2 Ethical and Legal Hurdles

From a legal standpoint, letting AI apply or revert production changes raises questions:

- What if an AI inadvertently violates **privacy regulations** or compliance rules?
- Who's liable for AI mistakes leading to data breaches?
- Complex cross-border data laws might hamper AI usage for test data provisioning or global deployments.

12.5.3 Trust and Cultural Adoption

Even with proven success, some organizations or stakeholders remain uneasy about AI's control:

- They might demand **full** human sign-off on every change.
- Corporate inertia, fear of job displacement, or legacy systems can slow AI adoption.
- Bridging that cultural gap requires transparent success stories, training, and robust governance.

12.6 Is Complete NoOps Truly Attainable?

12.6.1 The Last Mile of Human Judgment

Realistically, **100%** NoOps—where no human ever touches operations—may always be out of reach, because

- Novel or catastrophic incidents arise that AI can't handle with existing models
- High-stakes compliance or business decisions warrant human sign-off
- Evolution in frameworks or business priorities require human creativity and cross-domain thinking

*The ideal NoOps model may not eliminate humans, but rather shift their role to **high-level strategy, governance, and AI system oversight**.*

—McKinsey & Gartner AI Readiness Frameworks

NoOps is best viewed as an **aspirational horizon**: a state where 90–95% of routine ops is automated, letting humans handle truly unique or strategic challenges.

12.6.2 The Ongoing Collaboration

Even in advanced AI shops, humans remain crucial for

- **Policy creation** and oversight
- **Ethical governance** of data usage and privacy
- **High-level architecture** and user experience design
- **Innovation**—imagining new features or products AI alone wouldn't conceive

In essence, NoOps lowers operational friction so we can invest more human energy in **value creation**.

12.7 Chapter Summary and Conclusion

1. **Future AI: Multiagent and Human–AI Synergy**

    - Generative AI expands from code generation to entire **feature** creation, with agentic systems collaborating across DevOps.
    - Humans define goals, and AI handles day-to-day tasks, bridging coding, infra, pipeline, and advanced testing.

2. **Evolving Roles and Workforce**
   - Operators become automation architects, developers focus on logic and user experience, and QA shapes test strategy.
   - Leaner teams can manage larger systems, but require new AI and policy skills.

3. **Business Advantages**
   - Ultra-fast releases, dynamic user experiences, and lower ops overhead.
   - Real-time data feedback loops open new revenue possibilities or personalization strategies.

4. **Challenges and Governance**
   - Complexity of multiagent orchestration, trust building, ethical implications, and legal accountability.
   - Not all tasks can be automated—**human oversight** remains indispensable for novel scenarios and ethical considerations.

5. **The NoOps Horizon**
   - Complete autonomy for routine tasks is achievable, but expect a permanent **human–AI partnership** for strategic decisions.
   - The vision is software development moving at machine speed while humans innovate at strategic heights.

In closing, the era of **AI-driven DevOps** heralds unprecedented efficiency and agility. By leveraging multiagent systems, adopting a "stay in the flow" approach with NLP-driven IDE interactions, and establishing trust via strong oversight and governance, organizations can approach NoOps—freeing creative minds to focus on what truly matters: delivering remarkable products and experiences in a rapidly evolving digital world. The seeds are already planted; it's up to today's pioneers to cultivate a new generation of DevOps that merges human ingenuity with AI's relentless precision.

12.8 Conclusion and Final Thoughts

We've traversed a landscape where **DevOps**—originally the art of merging development and operations—has evolved into something far more expansive and intelligent:

- **Generative AI** transforming coding and testing
- **AI-driven IaC** for infrastructure and data provisioning
- **Adaptive CI/CD** that orchestrates deployments in real time
- **Multiagent systems** converging to push us toward a **NoOps** horizon

This final section synthesizes the book's core insights and offers a **call to action** for organizations and practitioners ready to embrace AI's full potential in DevOps.

12.9 Recap of the Journey

1. **DevOps Foundations**
   - Born out of siloed development and operations.
   - Showed how collaboration, continuous delivery, and automated pipelines accelerate software release cycles.

2. **Fragmented Ecosystems and Standardization**
   - Tool proliferation and data silos hinder efficiency.
   - Standardizing platforms, processes, and data is crucial before layering AI.

3. **Generative AI in Coding and Testing**
   - AI assists developers via tools like GitHub Copilot, speeds up boilerplate tasks, auto-suggests refactors, and even generates unit tests.
   - AI functional test suites self-heal, adapt to UI changes, and reduce QA overhead.

4. **Cloud-Native and Data-Centric**
   - Moving to microservices, containers, IaC, and integrated observability sets the stage for AI readiness.
   - Data provisioning for test or staging can also be AI-assisted, ensuring consistent, masked datasets.

5. **AI-Orchestrated CI/CD**
   - Pipelines become adaptive—intelligently selecting tests, scheduling canary rollouts, and auto-rolling back if anomalies arise.
   - NLP commands in the IDE let developers deploy or revert with minimal friction.

6. **Multiagent Systems and NoOps**
   - Specialized AI agents collaborate—coding, testing, infra, security, pipeline.

- Humans define policies, the system auto-manages, cutting operational toil while upholding compliance and best practices.

7. **Human–AI Collaboration**

    - Roles evolve, trust is built via audits and guardrails, and the organization shifts culturally to treat AI as a teammate.
    - NoOps doesn't remove humans; it elevates them to more strategic problem-solving.

12.10 Why It Matters Now

Software's pace is only quickening. Customers demand rapid feature rollouts, near-zero downtime, and strong security. Traditional manual processes can't keep up. AI offers

- **Speed and Scale**: AI can handle hundreds of daily tasks in parallel, from test updates to environment provisioning.
- **Consistency and Quality**: Less chance for human error, continuous scanning for compliance, and auto-rollback on anomalies.
- **Innovation Focus**: Devs and ops can do more creative design, user research, and product experiments.

As markets get more competitive, the ability to deliver new features **fast and reliably** is often the difference between leading or lagging. AI-driven DevOps is a major edge.

12.11 A Practical Call to Action

1. **Assess Your Current DevOps Maturity**

    - Identify the biggest pain points: fragmented toolchains, slow pipelines, test flakiness, or ops overload.

    - Prioritize which AI solutions (coding assistance, functional testing, IaC generation) could alleviate those pains.

2. **Standardize and Integrate**

    - Consolidate your tool stack where feasible—unified data flows and consistent naming conventions.

    - Ensure your code, tests, infra, and data provisioning are in version control, ready for AI oversight.

3. **Start Small with AI**

    - Introduce **GitHub Copilot** for coding or an **AI test generator**. Collect quick wins and build trust.

    - If comfortable, explore AI for infra scripts or partial pipeline orchestration (e.g., test selection).

    - Evaluate policy-as-code solutions to safely govern AI's changes.

4. **Encourage a Human-in-the-Loop Culture**

    - Keep initial AI outputs in "proposal" mode—require human review.

    - Track accuracy and build confidence; eventually automate low-risk tasks.

CHAPTER 12   THE FUTURE OF SOFTWARE DEVELOPMENT

5. **Upskill Teams**

    - Provide training on AI usage, prompt engineering, and policy definition.

    - Appoint "AI champions" who share success stories and guide best practices across squads.

6. **Incrementally Expand Autonomy**

    - Over time, let AI auto-apply more changes (like drift fixes or partial deployments) once proven safe.

    - Evaluate the ROI, watch for friction or hidden risks, and adjust guardrails as needed.

7. **Embrace NLP in the IDE**

    - If your AI stack supports it, enable chat or command panels so developers can request environment changes, data refreshes, or pipeline tasks with plain English.

    - Reduce context switching; keep feedback loops tight.

12.12 The Ongoing Evolution

NoOps is best seen as a **journey**, not an end state. Each incremental step—AI coding suggestions, AI testing, AI pipeline orchestration—brings significant efficiency gains and frees humans from repetitive tasks. The key is

- **Balance** between automation and oversight

- **Collaboration** among dev, ops, QA, and security teams on rules and guidelines

- **Steady iteration** to refine AI's capabilities as your organization's needs grow and shift

CHAPTER 12   THE FUTURE OF SOFTWARE DEVELOPMENT

Even as AI technology matures, humans remain essential for **strategic thinking, domain insight, and ethical governance**. The synergy of human creativity plus AI's relentless execution leads to a more vibrant, innovative, and resilient software delivery process.

12.13 Final Reflections

The future of software development belongs to those who effectively blend **human ingenuity** with **AI automation**. DevOps was already a revolution—unifying development and operations, cutting release cycles. Now, **AI** pushes us further, automating routine coding, testing, infra, and deployment tasks, enabling a **"stay in the flow"** developer experience.

Your challenge is to

- **Adopt AI** where it brings immediate wins, building trust step by step
- Invest in **training** and **policy** to ensure safe and effective usage
- **Continuously adapt** your processes and roles so that humans and AI complement each other
- Maintain a forward-looking mindset—new AI breakthroughs and frameworks appear rapidly, and early adopters often reap the competitive advantage

NoOps is not about removing humans from the loop—it's about **freeing** them to excel at the creative, strategic, and human aspects of software delivery. By embracing AI's potential and forging a culture of collaboration between people and machines, your organization can deliver software faster, safer, and more innovatively than ever before.

CHAPTER 12   THE FUTURE OF SOFTWARE DEVELOPMENT

Now is the time to start or accelerate your journey—transforming DevOps into an AI-driven force that redefines how we build and run software. The path to NoOps, while ambitious, offers unmatched rewards in efficiency, resilience, and the freedom to innovate. May your DevOps teams—and your AI agents—thrive together in this new era of intelligent automation.

12.14 Glossary—Part III

- **Multiagent AI**: A constellation of specialized agents (coding, testing, infra, security, orchestration) that coordinate software delivery end to end.

- **Autonomous Agent**: An AI process empowered to plan and execute changes (e.g., roll back a bad release) within predefined guardrails.

- **NoOps**: The aspirational state where 90%+ of operational tasks (build, test, scale, patch) run without human touch, leaving people to strategic work.

- **Human-in-the-Loop**: Oversight model in which humans review, approve, or override AI proposals, gradually expanding autonomy as trust grows.

- **Policy As Code**: Declarative rules (usually written for **Open Policy Agent (OPA)** or **Cedar**) that every AI agent must pass before merging or acting.

- **AI Guild/Tiger Team**: Cross-functional task force that pilots AI tools, curates prompt libraries, and tracks adoption KPIs.

- **AI-Accepted LOC**: Telemetry metric counting lines of code the team merged unedited after an AI suggestion.

- **Self-Healed Tests (Metric):** Percentage of functional tests automatically updated by an AI agent and later accepted.

- **Drift Auto-Remediation (Metric):** Share of infrastructure drifts patched automatically within a set SLA.

- **OpenAI Operator:** Experimental agent that interacts with an app exactly like a human through a built-in browser, validating full user journeys.

- **RunDeck/PagerDuty:** Ops tools that an AI "run-book copilot" can invoke to execute safe automations or page on-call engineers.

- **ChatOps:** Operations tasks handled directly in chat tools (Slack, Teams) where AI agents post status, metrics, and remediation options.

- **"Stay in the Flow":** Design principle: developers issue natural-language commands from their editor and receive instant AI feedback, never leaving their creative zone.

12.15 Bibliography

1. **Accelerate: The Science of Lean Software and DevOps** - Nicole Forsgren, Jez Humble, Gene Kim. IT Revolution Press, 2018.

2. **DORA (DevOps Research & Assessment).** *State of DevOps Reports (2024).* Google Cloud.

3. **Puppet.** *State of DevOps Report (2023).*

4. **GitHub Copilot & AI Pair Programming.** OpenAI & Microsoft, 2024.

5. **Google Research.** *AI Developer Productivity Report (2024).*

6. **McKinsey & Company.** *AI-Driven Software Development & Code Refactoring Trends (2024).*

7. **JP Morgan Chase AI Testing Initiative.** *Case Study on AI-Generated Test Cases, 2023.*

8. **Diffblue Cover.** *Automated Java Test Generation with AI, 2024.*

9. **Tricentis.** *AI Test Automation & DevOps Efficiency Report, 2024.*

10. **Google Site Reliability Engineering (SRE) Principles.** *The Site Reliability Workbook.* O'Reilly Media, 2018.

11. **Dynatrace, Moogsoft, and Splunk.** *AIOps in DevOps & IT Operations, 2024.*

12. **AWS DevOps Guru & Google Cloud Autopilot.** *Cloud-Based AI for DevOps Optimization, 2024.*

13. **Firefly AI for Infrastructure Drift Detection.** *AI for Self-Healing Infrastructure, 2024.*

14. **Quali's Torque AI for Automated Provisioning.** *AI-Generated Terraform & Infrastructure as Code, 2024.*

15. **Netflix.** *Chaos Engineering & AI-Driven DevOps, 2024.*

16. **Meta (Facebook).** *AI-Optimized Data Center Operations, 2024.*

17. **Microsoft Security Copilot.** *Machine Learning for Automated Security Analysis, 2024.*

18. **Amazon CodeGuru.** *AI-Driven Code Review & Optimization, 2024.*

19. **Digital.ai DevOps Governance Reports.** *Best Practices for AI-Integrated DevOps Pipelines, 2024.*

20. **Gartner Hype Cycle for AI in Software Development (2024).**

21. **Futurum Research.** *Agentic AI & Autonomous DevOps, 2025.*

22. **MIT & Gartner Reports.** *Multi-Agent AI Systems in DevOps, 2025.*

23. **McKinsey & Gartner AI Readiness Frameworks.** *Enterprise Adoption of AI in DevOps, 2024.*

24. **DORA's 2024 DevOps & AI Survey.** *AI Adoption, Trust, and Performance Metrics in DevOps.*

25. **State of DevOps Toolchain Survey (2016).** *The Impact of Tool Sprawl on DevOps Efficiency.*

26. **Capital One.** *Case Study: Standardization & Automation in DevOps Transformation, 2023.*

Index

A

Adaptive CI/CD, 221
Adaptive pipelines, 151
Agentic AI—multiple specialized agents collaboration, 213
Agile methodologies, 3, 4
Agile principles, 19
Agile projects, 5
AI-accepted code lines, 166
AI assistants, 166
AI automation, 226
AI autonomy, 188, 204, 208
AI coding, 225
    assistants, 106, 107
    best practices, 114, 115
AI-driven automation, 18, 39, 50, 67
AI-driven development
    code suggestions, 115
    developer experience, 116
    NoOps, 116
AI-driven DevOps, 13–17, 32, 37, 49, 209, 216, 220, 223
AI-driven functional testing
    benefits, 124
    best practices, 125, 126
    challenges/caveats, 124
    end-to-end validation, 122
    Functionaize, 121
AI-driven IaC, 136–138, 141–142, 221
AI-driven testing tools, 119, 122
AI-enhanced testing workflows
    generating tests, 122, 123
    integration testing, 123
    self-healing, 123
AI-generated code, 14, 114, 165
AI-generated infrastructure, 164
AI-orchestrated CI/CD, 222
    agents and NoOps
        multiagent pipeline collaboration, 160
        NLP-driven flow from IDE, 161
        real-time observations and automated fixes, 160, 161
    AI-driven pipeline optimization
        intelligent test selection, 152, 153
        partial/on-demand deployment sequences, 153
    defining risk profiles, 157
    deploy and release strategy optimization
        blue-green, 155
        canary, 155

INDEX

AI-orchestrated CI/CD (*cont.*)
    real-time telemetry feedback, 155
    rolling deploy, 155
    Ecommerce company's AI-managed pipeline, 158, 159
    guardrails and policy, 157
    incremental adoption, 158
    integrate observability, 158
    train the AI, 157
    predictive failure analysis and remediation
        anomaly detection, 154
        auto-apply fixes/reruns, 154
    smarter pipelines
        complexity and staging bottlenecks, 150
        real-time feedback *vs.* blind scripts, 151
    stay in the flow, 156
AI readiness, 50
    AI platforms and ML services, 65
    automation tools, 43
    code style guidelines, 47
    organization, 47
    pipelines and developer environments, 43
    production incidents, 47
    subset, 47
AI testing
    autonomous test agents, 128
    Dev/QA/Ops, 128
    OpenAI Operator, 129

Anomaly detection, 32, 43, 60, 64, 65, 121, 154, 158, 161, 189, 191
Automated AIOps, 12
Automated Rollout, 62
Automated testing, 8, 61
Automation scripts, 30
Autonomous agents, 43, 48, 52, 86, 183, 227
Autonomous code generation
    code as conversation, 213, 214
    interactive agents in the IDE, 214
Autonomous orchestration, 161
Autonomous scaling, 65
Auto-remediated infra drifts, 166

B

Business strategy
    data monetization and AI feedback loops, 216, 217
    time to market and continuous innovation, 216

C

Chaotic fragmentation, 27, 50
CI/CD agent, 180, 185–187, 190
Cloud-based IDEs, 31, 46
Cloud environments, 53, 66
Cloud-native, 53
    architectures, 50, 52, 53, 65
    environment, 60

pipeline, 62
technologies, 64
Cloud Ops, 29
CodeQL, 91, 93, 96, 97, 176, 180
Coding agent, 184–186, 204, 213, 214
Cognitive load, 23, 28, 88, 110
Collaboration friction, 26–27, 35
Communication and decision-making, 185, 194
Containerization, 54–56, 100
Containers, 29, 50, 53, 55–56, 59, 88, 90, 100, 133, 222
Context switching, 23, 26–27, 31, 36, 41, 141, 156, 159, 190, 195, 225
Copilot-like chat/status panels, 156
Costs drop, 165
Cumulative friction, 28

D

Data-centric design, 64, 72
Data-centric approach, 53–69
Data-driven adaptive approach, 155
Data management and multiagent infrastructure, 144, 145
DataOps, 11–12, 99
Data privacy, 134, 188
Data provisioning
and IaC (*see* Infrastructure as code (IaC))
Data silos, 18, 24–25, 35, 61, 222

Data unification, 47, 51
Dedicated AI agents, 160
DevOps, 6, 9, 10, 20–22, 25, 26, 29, 30, 33–35, 37, 42, 50, 52, 54, 67, 221
ADKAR, 15
Agile, 5
agility and reliability, 8
AI benefits, 15
aims, 6
AI post-mortems, 15
change-management, 16
CI/CD pipelines, 8, 12
and cloud-native technologies, 9
culture, 41
delivery pipelines, 6
DevSecOps, 11
DORA, 7
dysfunctions, 5
elite performers and traditional organizations, 7
evolution, 13, 17
foundational themes, 6
foundations, 221
friction, 4
initiatives, 14
innovation, 43
lifecycle stage, 10
MLOps, 11
model, 4
monolithic architectures, 10
NoOps, 12
operations, 4

INDEX

DevOps (*cont.*)
    origin, 17, 19
    Patrick Debois, 5
    real-time feedback, 60
    software, 10, 18
    solutions, 32
    toolchain, 12
    transformation, 8
DevOps reference architecture
    agentic AI, 85
    AI-driven test orchestration, 85
    consistent guardrails, 86
    end-to-end integration, 72
    generative AI, 85
    hallmark indicators, 80, 81
    organizational design, 79, 80
    pitfalls, 83, 84
    reference model
        artifact management, 75
        automated build and test (CI), 75
        continuous delivery/deployment (CD), 76
        IaC, 75
        IDEs, 74
        integration/system testing, 75
        observability and Ops, 76
        platform-agnostic analytics layer, 76
        requirements and planning, 74
        version control and code collaboration, 74

SaaS, 82, 83
security and compliance, 73
self-healing, 73
self-service, 73
single source of data, 72, 73
workflow, 76, 78
DevSecOps, 62
    security, 11
    tools, 67
Duplicate data entry, 26

E

Ecommerce company's AI-managed pipeline
    canary deploy and observability, 159
    NLP commands in IDE, 159
    results, 159
    selective test execution, 158
Ecommerce, Functionaize, 126, 127
End-to-end telemetry, 164
End-to-end test suite, 126
Executives, 88

F

Fast feedback loops, 27
Feedback, 62
FinTech startup
    initial setup, 143
    microservices, 142
    outcomes, 144
    remediation and scaling, 143, 144
    review and integration, 143

Fragmentation, 32, 33, 35
Functionaize, 163, 164, 166, 170, 172, 174–177
Functional tests, 118, 120–122, 129, 130, 133, 228

GitHub Copilot operates, 164
GitHub Enterprise Cloud, 89
"Golden Pipeline" approach, 32, 40, 45–46, 51, 79, 84, 89, 100
GPT-4 model, 143

G

Generative AI, 149, 161, 163, 164, 167, 172, 180, 183, 221, 222
   challenges, 113
   data provisioning (*see* Data provisioning)
   IaC (*see* Infrastructure as code (IaC))
   limitations, 113
   productivity and code quality, 109–111
   workflow
      code reviews, 109
      documentation, 108
      edge cases, 108
      GitHub Copilot, 108
      team collaboration, 109
GHAS, *see* GitHub Advanced Security (GHAS)
GitHub, 48, 91
GitHub Actions, 27, 44, 49, 75, 82, 90, 96–97, 168, 174, 179
GitHub Advanced Security (GHAS), 87, 91, 92, 168, 169, 171–174, 176, 177
GitHub Copilot, 13, 163, 164, 174, 175, 189, 214

H

Human–AI collaboration, 223
   auditable actions and policy checks, 201
   cultural and organizational shifts
      continuous iteration on roles and processes, 206
      embrace AI as a teammate, 205, 206
   developers as product creators, 199
   ethical and compliance considerations
      bias and reliability, 204
      boundaries of AI autonomy, 204
      legal and accountability, 205
   human touch, 207
   lifelong learning and evolve AI, 207
   long-term NoOps vision, 206, 207
   from manual operators to automation architects, 198
   security and compliance roles, 200

Human–AI collaboration (*cont.*)
  transparency and explainability, 202
  upskilling and team dynamics
    collaboration with AI agents, 203
    training developers and Ops, 202
    value-driven roles, 203
Human-in-the-loop, 200

I

IaC, *see* Infrastructure as code (IaC)
IDE, *see* Integrated Development Environment (IDE)
Infrastructure agent, 184–186
Infrastructure as code (IaC), 57, 75, 90, 100
  CI/CD pipelines, 58
  and data management
    clear guardrails, 141, 142
    continuous learning, 142
    cross-functional collaboration, 142
    human oversight, 141
    security, policy, and data compliance checks, 141
  and data provisioning
    automated script creation, 137
    complexity and rapid changes, 134, 135
    and data management, 134

DevOps and IDE NLP, 135
drift and misconfiguration remediation, 139
predictive scaling, 138, 139
pull request and CI/CD gates, 135
refactoring and modernization, 138
and test data, 133
workflow, 136
DevOps, 58
Git repos, 57
GPT, 170, 174
HashiCorp, 57
IDE-centric, 140, 141
NLP-driven actions, 140, 141
Pulumi, 57
Integrated AI-powered DevOps, 189
Integrated Development Environment (IDE), 30, 36, 134, 135, 137, 140, 141
  cloud-based, 31
  code and unit tests, 164
  ecosystems, 31
  and scattered plug-ins create friction, 31
  NLP-driven flow in, 145, 146
Integration Overhead, 23
Integration tests, 75, 77, 119–131

J

JetBrains alternative, 49

INDEX

K

KPIs
  adoption and business, 172, 173
  and success metrics, 177, 178
Kubernetes, 56, 66, 89

L

Large language models (LLMs), 12, 105–107, 163, 179
LLMs, *see* Large language models (LLMs)

M

Manual data transfer, 28
Masking library, 143
Mean time to recovery (MTTR), 25, 61, 67, 76, 88–90, 97, 98, 168, 172, 178
Microservices, 46, 50, 134
  concept, 55
  model, 55
Minimal context switching, 190
Mitigation playbook
  adoption and business KPIs, 172
  AI-assisted coding and testing
    Copilot coverage mandate, 169
    prompt library and style guide, 169
    self-healing test, 170
  drift, lighthouse squads and policy, 94
  infrastructure and operations
    AI IaC generator, 170
    predictive scaling agent, 170
    run-book Copilot, 171
  pillars, 92
  platform guardrails
    centralize AI telemetry, 168
    lock in the single IDE (VS code), 168
    policy-broker bot, 168, 169
  repos and pipelines, 92
  security and compliance, 93
    AI-aware SBOM and license scan, 171
    continuous policy drift audit, 171
    IDE-level secret push protection, 171
  telemetry, 92, 93
  VS Code workspace, 94
  wire speed/security/spend, 93
  workspace, 93
ML-based solutions, 60
Monolithic applications, 54
MTTR, *see* Mean time to recovery (MTTR)
Multiagent infrastructure and data management, 144, 145
Multiagent NoOps, 189
  era, 163
  practices for embrace, 191, 192

237

INDEX

Multiagent systems, 221, 222
    benefits, 187, 188
    challenges, 188

N

Netflix, 8, 17, 19, 229
NLP, 134, 135, 137, 166, 190, 199, 202
NLP-Driven Flow in IDE, 145, 146
No Operations (NoOps), 12, 13, 20, 37, 60, 66, 183, 186
    automation, 48
    aspirational horizon, 219
    challenges and limitations
        complexity and interagent conflicts, 217
        ethical and legal hurdles, 217
        trust and cultural adoption, 218
    collaboration, 219
    human judgment, 218
    infrastructure and test data, 146
    paradigm, 212
    value creation, 219
    vision, 130
NoOps, *see* No Operations (NoOps)
NoOps vision, 130, 192, 193
    constant evolution, 193
    continued role for humans, 193

O

Observability, 58–61, 76, 101, 159, 162, 191
OPA, *see* Open Policy Agent (OPA)
OpenAI Operator, 129–131, 145, 228
Open Policy Agent (OPA), 141, 191, 201, 227
OpenTelemetry, 89, 96, 101, 168
Operator-like approaches, 160
Opsera, 87–90, 166, 172
    KPIs and success metrics, 177, 178
    quick-start checklist
        AI guild" tiger Team, 173
        baseline AI readiness, 173
        connect AI agents, 174
        lock the workspace, 173
        stand-up the policy broker, 174
        sync with Opsera, 174
        wire telemetry for AI, 174
    sequenced migration plan
        code and test expansion (weeks 7–10), 175, 176
        infrastructure and operations (weeks 11–14), 176
        org-wide security and policy (weeks 15–18), 176, 177
        pilot service (weeks 3–6), 175
        proof of concept (weeks 0–2), 175
        scale and optimize (weeks 19–24), 177
Opsera Unified Insights, 164, 173, 175

P, Q

Paved IDE, 167
Perform load tests, 60
Platform engineering, 30, 31, 33, 37, 45, 46, 48, 51, 88, 89, 203, 215
"Platform First" approach, 48–49
Platform synergy, 107, 117
Policy broker, 167, 170, 172–175
Policy-broker bot, 168, 169
Predictive scaling bot, 164, 174, 176
Prompt library, 169, 173, 176, 178

R

Real-time feedback, 60, 67, 68, 88, 156
Real-time feedback *vs.* blind scripts, 151
Real-time visibility, 7
Repeatable playbook
   checklist, 95, 96
   KPIs, 98
   metrics, 98
   migration plan, 96, 97
Repetitive tasks, 6, 28, 225
ROI, 10, 164, 225

S

SaaS, *see* Software-as-a-Service (SaaS)
Scripted approach, 155
Security agent, 186
Security and Policy Guardrails, 165
Security shifts, 165
Security vulnerabilities, 25, 28
Self-healing tests, 165, 166, 170
"Shift-left" security, 10, 89, 97, 101
Siloed data, 21, 25
Simian Army, 8
Software-as-a-Service (SaaS), 26, 82, 83, 88, 90
Software delivery, 163, 226, 227
Software's pace, 223
Spark innovation, 27
Specialized AI agent
   CI/CD orchestration agent, 184
   coding agent, 184
   functional testing agent, 184
   infrastructure agent, 184
   operator-like system agent, 185
Split the monolith, 63
Standardization, 35, 37, 40, 43, 44, 49, 50, 52, 68
   center of excellence/lighthouse projects, 44
   CI/CD, 40
   DevOps context, 40
   DevOps efforts, 41
   fundamental toolchain, 41
   IDEs, 40
   security and compliance, 42
   toolkit, 42
Standardized metrics, 61, 69
Stay in the flow, 134, 135, 140, 141, 145
   AI-orchestrated CI/CD, 161

Stay in the flow (*cont.*)
    IDE-centric, 156
    natural language triggers, 156
    NLP-driven CI/CD control, 156
    quick feedback and reduced context switching, 156
System and integration testing, 119–131, 164

T

Teams experiment, 7
Terraform, 22, 77, 82, 89, 100, 134, 137, 142–144
Terraform modules, 42, 82, 89, 137
The Phoenix Project, 5–6
Tool-based silos, 25
Toolchain, 23, 50
Tool Sprawl, 32
    developers and operators, 23
    discrete tools, 22
    platforms, 22
    staff turnover, 23
    toolset, 23
Tool tax, 23, 36, 51

U

Unified insights, 87–90, 93, 163, 165, 168

Unified interface, 190
Unit tests, 111
    benefits, 112
    best practices, 114, 115
    GitHub Copilot, 112
    individual functions/classes, 120
    limitations, 112

V

Velocity, 88, 90, 164, 168
Velocity Conference, 5
Virtual machines (VMs), 55–56
Visual Studio Code (VS Code), 30, 87–89, 91, 93, 94, 96, 164, 168, 190
VMs, *see* Virtual machines (VMs)
VS Code, *see* Visual Studio Code (VS Code)
Vulnerability, 113, 154

W, X, Y, Z

Workforce and organizational impact
    leaner teams, faster delivery, 215
    upskilling and new roles, 215

GPSR Compliance

The European Union's (EU) General Product Safety Regulation (GPSR) is a set of rules that requires consumer products to be safe and our obligations to ensure this.

If you have any concerns about our products, you can contact us on

ProductSafety@springernature.com

In case Publisher is established outside the EU, the EU authorized representative is:

Springer Nature Customer Service Center GmbH
Europaplatz 3
69115 Heidelberg, Germany

www.ingramcontent.com/pod-product-compliance
Lightning Source LLC
LaVergne TN
LVHW021957060526
838201LV00048B/1596